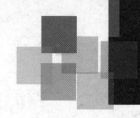

计算机数字控制技术

Computer Numerical Control Technology

主　编　赵　巍　董靖川
副主编　赵　楠　刘鹏鑫　张翔宇
参　编　[巴] MALI MANSOOR (阿里)
　　　　[巴] MUHAMMAD KASHIF IJAZ (卡什夫)
　　　　[巴] RAMEEZ UL HASSAN (哈森)

清华大学出版社
北京

内 容 简 介

本教材按照数控机床的主要工作流程编写,详略得当;重运用、强实践,将抽象的原理与实际应用有机结合;追踪先进技术的发展,涵盖智能制造技术最新成果。关键词语采用双语说明方式。

本教材共有5章,包括数控机床发展史、数控系统工作原理、伺服系统结构及工作原理、机械结构、智能制造等内容。教材还包含实验指导书,强化对学生实践创新能力的培养。

版权所有,侵权必究。侵权举报电话: 010-62782989　13701121933

图书在版编目(CIP)数据

计算机数字控制技术:英文/赵巍,董靖川主编. —北京:清华大学出版社,2019
ISBN 978-7-302-52368-0

Ⅰ. ①计… Ⅱ. ①赵… ②董… Ⅲ. ①数控机床－数控系统－高等学校－教材－英文 Ⅳ. ①TG659

中国版本图书馆 CIP 数据核字(2019)第 039034 号

责任编辑:冯　昕　赵从棉
封面设计:傅瑞学
责任校对:刘玉霞
责任印制:杨　艳

出版发行:清华大学出版社
　　　　网　　址: http://www.tup.com.cn, http://www.wqbook.com
　　　　地　　址: 北京清华大学学研大厦 A 座　　邮　编: 100084
　　　　社 总 机: 010-62770175　　邮　购: 010-62786544
　　　　投稿与读者服务: 010-62776969, c-service@tup.tsinghua.edu.cn
　　　　质量反馈: 010-62772015, zhiliang@tup.tsinghua.edu.cn
印 装 者: 涿州市京南印刷厂
经　　销: 全国新华书店
开　　本: 185mm×260mm　　印　张: 10.75　　字　数: 259 千字
版　　次: 2019 年 6 月第 1 版　　印　次: 2019 年 6 月第 1 次印刷
定　　价: 49.00 元

产品编号:076562-01

前　言

中巴、中非友谊源远流长，加大教育国际援助力度，具有针对性地开展紧缺教学资源建设，不仅可以提高多层次、宽领域的国际教育水平，而且有助于增进国际友谊，为"一带一路"合作绘制"工笔画"。

数控技术作为先进制造领域的核心技术，是教育援外机械专业本科生和研究生的必修知识。然而，现有数控技术课程双语教学资源不足，现有教材出版日期较早，缺乏先进数控技术及应用等方面的知识。为填补本领域空白，有针对性地为发展中国家培养数控技术人才，特编写本教材。

本教材分为5章，主要涵盖数控机床发展史、数控系统工作原理、伺服系统结构及工作原理、机械结构、智能制造等内容。本教材按照数控机床的主要工作流程编写，详略得当；重运用、强实践，将抽象的原理与实际应用有机结合；追踪先进技术的发展，涵盖智能制造技术最新成果。关键词语采用双语说明方式。可以作为中巴、中非合作办学的机械工程专业留学生专业教材、高年级本科生双语教材，也可以供数控技术人员自学参考。教材包含实验指导书，强化学生实践创新能力的拓展。

本教材第1章由天津职业技术师范大学赵巍、旁遮普天津技术大学 MUHAMMAD ALI MANSOOR 编写；第2章由天津职业技术师范大学赵楠、旁遮普天津技术大学 MUHAMMAD KASHIF IJAZ 编写；第3章由天津大学董靖川编写；第4章由天津职业技术师范大学赵巍、张翔宇，旁遮普天津技术大学 RAMEEZ UL HASSAN 编写；第5章由天津职业技术师范大学刘鹏鑫、赵楠，湖州职业技术学院王英杰编写。全书由付晓燕校稿，旁遮普天津技术大学 Dr. Syed Asim Ali Shah、天津职业技术师范大学满佳老师、北京格瑞纳电子产品有限公司赵伟工程师以及研究生刘梦莹、王丽娜、何苗等也参与了本教材的审核、编辑等工作。

在教材的编写过程中参考了一些数控技术相关的教材和资料，特向其作者表示真诚的感谢。鉴于作者水平有限，书中不可避免地存在缺点、不足，敬请各位读者批评指正。

编　者
2019年1月

CONTENTS

Chapter 1   Introduction ·· 1

     1.1   The history of numerical control machine tools ·················· 1
     1.2   Basic principles ··· 2
           1.2.1   The principle of computer numerical control
                  (CNC) machine tools ····································· 2
           1.2.2   The definition of numerical control ····················· 3
     1.3   The characteristics of CNC machine tools ······················· 4
     1.4   Introduction to main CNC machine tools ························· 5
     1.5   The development of machinery manufacturing systems ········ 9

Chapter 2   Computer numerical control (CNC) systems ························· 13

     2.1   The working principle of the CNC system ····················· 13
           2.1.1   Development of the NC system ······················· 13
           2.1.2   The composition and working principle of the CNC system ······ 19
           2.1.3   The hardware of the CNC unit ······················· 21
           2.1.4   Hardware composition of CNC devices ············· 22
           2.1.5   Software components of the CNC system ·········· 24
           2.1.6   The functions of CNC devices ······················· 25
     2.2   Interpolation theory ·· 28
           2.2.1   Overview of the interpolation ························· 28
           2.2.2   Point by point comparison interpolation ············ 31
           2.2.3   Digital differential analyzer interpolation ·········· 37
           2.2.4   Sampled data interpolation ···························· 41
           2.2.5   Acceleration and deceleration control ·············· 42
     2.3   Tool compensation principle ····································· 43
           2.3.1   Tool length compensation ····························· 44
           2.3.2   Tool radius compensation ····························· 46

Chapter 3   Servo systems ··· 51

     3.1   Overview of servo systems ······································ 51
           3.1.1   Servo system components ····························· 51

    3.1.2 Servo system classification ·· 52

    3.1.3 Basic requirements of servo systems ···································· 55

  3.2 Commonly used driving elements ··· 57

    3.2.1 Stepper motors ··· 57

    3.2.2 DC servo motors ·· 72

    3.2.3 AC servo motors ·· 79

    3.2.4 Linear motors ··· 90

  3.3 Commonly used detecting elements ··· 94

    3.3.1 Rotary encoders ··· 95

    3.3.2 Linear encoders ··· 99

Chapter 4 Mechanical structure of the CNC machine tool ························· 102

  4.1 Requirements for the CNC machine tool for mechanical structure ······· 102

  4.2 Current common types and layout of CNC machine tools ············· 107

    4.2.1 CNC Lathes ·· 107

    4.2.2 Machining centers ·· 109

  4.3 Main drive system and spindle components ································· 112

  4.4 CNC machine tool feed drive systems ·· 124

    4.4.1 Requirements for CNC machine tools on the

       feed drive system ··· 124

    4.4.2 Connection between the motor and the lead screw ·········· 126

    4.4.3 Ball screw nut pairs ·· 129

    4.4.4 Guide rail slider pairs ··· 135

  4.5 Self-made one-dimensional feed mechanism ································ 140

Chapter 5 Intelligent manufacturing ··· 142

  5.1 The development and application of intelligent manufacturing ··· 142

    5.1.1 American industrial Internet ··· 142

    5.1.2 Germany industry 4.0 ··· 143

    5.1.3 Made in China 2025 ·· 144

  5.2 The core of intelligent manufacturing ··· 146

  5.3 Technical support for intelligent manufacturing ·························· 147

  5.4 A case study of intelligent manufacturing—Siemens Amberg factory ······ 154

Experiment A Two-dimensional motion control platform ························ 156

Experiment B The measure of positioning accuracy and repeated positioning

       accuracy to one-dimensional motion mechanism ················· 160

Recommended book ·· 164

Chapter 1　Introduction

1.1　The history of numerical control machine tools
（数控机床史）
_{shù kòng jī chuáng shǐ}

To assist in engineering calculating, a computer was developed at the University of Pennsylvania in 1946. It was called Electrical Numerical Integrator and Calculator which consisted of huge mass of tubes and wires. Although it was slow and stood no comparison with modern computers, it was the best at that time.

The birth of Numerical Control (NC，数控_{shù kòng}) is generally credited to John T. Parsons, a machinist and salesman at his father's machining company, Parsons Corp. In 1942, *The United States Air Force* signed a contract with Parsons Corporation to build the wooden stringers (or ribs, see Figure 1-1) in the rotor blades (转子叶片_{zhuàn zǐ yè piàn})(only 17 points were given to define the outline of stringers), but

Figure 1-1　One kind of ribs

one of the blades failed. Parsons suggested a new method, which led him to consider the possibility of using stamped metal stringers instead of wood. A device would not be easy to produce given the complex outline. Looking for ideas, Parsons visited Wright Field to see Frank Stulen, the head of the Propeller Lab Rotary Wing Branch. Stulen's brother worked at Curtis Wright Propeller, and mentioned that they were using punched card calculators for engineering calculations. When Parsons saw what Stulen was doing with the punched card machines, he asked Stulen if they could be used to generate an outline with 200 points instead of the 17 they were given, and offset each point by the radius of a mill cutting tool. If you cut at each of those points, it would produce a relatively accurate cutout of the stringer(横梁_{héng liáng}). This could cut the tool steel and then easily be filed down to a smooth template for stamping(冲压_{chòng yā}) metal stringers.

Stulen made the program for the stringer which is comprised of large table consisting of numbers, without any problem. Later the program was communicated to machine. There are three operators: one reads the card while other two move the tool in

X-axis and Y-axis to follow the cutting profile. For each pair of numbers, the operators had to move the cutting head to the indicated spot and then lower the tool to make the cut. This was called the "by-the-numbers method"(数字控制技术), or more technically, "plunge-cutting positioning"(切入式定位). It was a labor-intensive prototype of today's 2.5-axis machining.

At that point, Parsons conceived of a fully automated machine tool. With enough points on the outline, no manual working would be needed to make it.

In 1949, the Air Force arranged funding for Parsons to build his machines, but he was confronted by the problem that had prevented convergence of Jacquard-type controls with machining.

In the spring of 1949, Parsons turned to Gordon S. Brown's Servomechanisms Laboratory at MIT, which was a world leader in mechanical computing and feedback systems(反馈系统).

A tripartite agreement was arranged between Parsons, MIT, and the Air Force. In March 1952, the MIT Labs held the first demonstration of the NC machine using a punched tape(穿孔带) to generate the movement of three axes. See Figure 1-2.

Figure 1-2　The first successful NC machine, developed by MIT

1.2　Basic principles

1.2.1　The principle of computer numerical control(CNC) machine tools

1. Preparing stage

Manufacturing data is determined by the parts drawing and technical requirements, including tool trace coordinates, spindle, feed, cutting depth, tool and so on. Other auxiliary data, for example, tool changing, workpiece clamping(夹紧 jiā jǐn) and release(松开 sōng kāi), cooling, lubrication, is defined, too.

2. Programming stage

For computer numerical control(CNC, 计算机数控 jì suàn jī shù kòng) machining, programs are written in forms of G-code and M-code, which are the understandable languages of CNC system. For instance, G00, G02, M03, M30 and so on.

3. Transmitting stage

The programs are transmitted into CNC system through the tape (outdated), disk, Internet and so on.

4. Machining stage

When the program runs, it translates codes into signals which drive the motors accordingly. The motors then move the workpiece and cutting tool to get desired dimensions required.

1.2.2 The definition of numerical control

Numerical control(NC, 数值控制,简称"数控" shù zhí kòng zhì / shù kòng) is also called digital control(数字控制 shù zì kòng zhì). NC technology is to use digital information to realize controlling automation. NC has been widely used in trajectory control of the mechanical movement and switching control of the mechanical system, such as robots, machine tools, production lines and so on. The objects controlled by digital information are varied, but the NC machine tools are the most typical NC equipment.

What are NC machine tools? They mean the use of digital signals to control the machine movement and process.

What are CNC machine tools? They are developed with microprocessors or special computers as NC systems. The control system carries out automatic machining of the workpiece. Today, all the machine control units are based on the computer technology.

In CNC machine tools, digital codes, including numbers, letters and symbols, are taken to express the workpiece dimensions and a variety of operations, such as setting spindle speed, loosening or fastening the cutter, setting feed rate, shutting on or off the coolant automatically, running or stopping the program, etc. And then digital codes are transmitted into the NC device through the control medium (disks, serial ports, network). The NC device deals with the input information and performs calculation, then sends a corresponding control signal to the servo system or other driving components for automatic machining the workpieces. The CNC structure is shown as Figure 1-3.

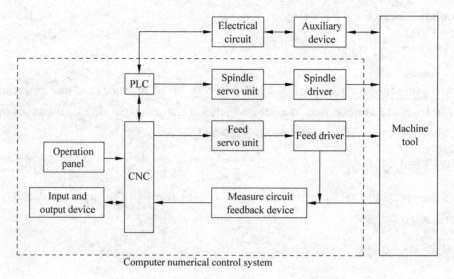

Figure 1-3　CNC system structure

1.3　The characteristics of CNC machine tools

1. Adaptability

Adaptability also means flexibility, which refers to that the CNC machine can adapt to different jobs very well. For every new part, we only need to rewrite the program and enter the new one to complete the machining of the new part without changing the mechanical part and the control part. The production process is done automatically. This provides a great deal of convenience for single part, small batch production and trial production of complex structural parts. Strong adaptability is the most prominent advantage of CNC machine tools, which resulted in its rapid development.

2. High precision and stability

Firstly, compared with manual machining method, CNC reduces or eliminates operator's influences. Secondly, during the design and manufacture of CNC machine tools, a number of measures are taken to achieve high accuracy and stiffness for the mechanical parts and structure. The pulse equivalency normally reaches 0.0001—0.01 mm, and CNC machining accuracy has varied from the past ± 0.01 mm to ± 0.001 mm or even higher due to the compensation of the chain reverse gap and screw pitch error realized by the NC device. In addition, the stiffness and thermal stability of transmission system and the machine structure are higher for the CNC machine tool. In particular, the batch production consistency is improved remarkably, so high product qualification and stability is achieved.

3. Higher automation, lower labor intensity

CNC machine tool operations can be carried out automatically by the control program. The operator only needs to set the job and tool and the rest is done by the program. The operator just observes the process and performs inspection, which has greatly reduced the labor. In addition, instead of being open for the conventional machine, CNC machine tool is generally closed during processing, which are both clean and safe. Today CNC machines are available in esthetic colors ranging from green to gray or white.

4. Higher efficiency, shorter machining and auxiliary time

Parts processing time mainly consists of machining time and auxiliary time. CNC machine tools are equipped with wide range of spindle speeds and feeds, so for each process we can choose the most appropriate setting. Due to the rigidity of the CNC machine tool structure, it is allowed to carry out the strong cutting, which improves the cutting efficiency and saves the machining time. Auxiliary time is also reduced as compared with the conventional machine, for example, fast movements and shorter workpiece clamping time result in much shorter automatic tool change time.

In addition, an automatic tool changer made it possible to carryout multi-order continuous processing without any delay.

5. Being conducive to the modernization of management

In the CNC machine tools, digital information and standard codes are used and transmitted, which lays the foundation for computer-aided design/manufacturing and management integration.

1.4　Introduction to main CNC machine tools

With regard to the manufacturing process, CNC machine tools include CNC lathes, CNC milling machines, machine enters, CNC drilling machines, CNC boring machines, CNC grinding machines, CNC electric discharge machines, CNC wire-cut electric discharge machines, CNC laser beam machines, CNC punching machines or punching presses, CNC ultrasonic machines, CNC gear holling machines, CNC plasma cutting machines, CNC bending machines, CNC water cutting machines, parallel machine tools, coordinate measuring machines and so on.

The most common CNC machine tools are introduced as follows.

1. CNC lathes(数控车床)

A CNC lathe (see Figure 1-4) includes spindle(主轴), slide plate(溜板), toolholder(刀架) and so on. CNC system, including LCD panel(液晶显示屏), control panel(控制面板), electrical control system(电气控制系统).

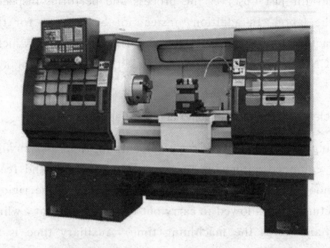

Figure 1-4　CNC lathe

CNC lathe generally has two-axis interpolation function. Z-axis is parallel to the direction of the spindle axis while X-axis can move vertically to the spindle. The latest turning and milling center with a C-axis can turn and mill the workpiece automatically with the milling cutter placed in the tool turret.

The CNC lathe is mainly used to process shaft or disc parts, such as cutting the inner and outer cylindrical surface, cone surface, thread surface, end surface, groove, chamfering, etc. For the rotary bodies, drilling, reaming and boring can also be done.

2. CNC milling machines(数控铣床)

CNC milling machines (see Figure 1-5) can process three-dimensional complex surface. They are widely used in the automotive, aerospace(航空航天), mold(模具) and other industries. They can be divided into vertical CNC milling machines, horizontal CNC milling machines, duplicating CNC milling machines.

3. Machining centers(加工中心)

The machining center (see Figure 1-6) results from the development of CNC machine tools, generally considered machining centers for the CNC boring and milling machines with an automatic tool controller (ATC). Machining center can do milling(铣削),

Figure 1-5　CNC milling machine

boring, drilling, broaching, fraising, tapping and other processing. Machining centers are classified as vertical ones and horizontal ones. The former ones' spindle axes are vertical, the latter ones' spindle axes are horizontal. The horizontal machining center is particularly used to machine the large, boxy and heavy workpieces.

(a)　　　　　　　　　　　　(b)

Figure 1-6　Machining center

(a) Vertical machining center; (b) Horizontal machining center

4. CNC drilling machines（数控钻床）

CNC drilling machines (see Figure 1-7) can be divided into vertical ones and horizontal ones. Both of them mainly used for drilling(钻削)and tapping, but also for simple milling.

5. CNC grinding machines(数控磨床)

CNC grinding machines are mainly used to machine hard surfaces with high precision. They can be classified as CNC surface grinders, inner cylinder grinding machines, contour grinding machines and so on. With the development of the automatic grinding wheel compensation technology, automatic grinding wheel(自动砂轮) dressing technology and grinding fixed cycle technology, CNC grinding function becomes increasingly strong.

6. CNC electrical discharge machines(EDM) (数控电火花机床)

Figure 1-7 CNC drilling machine

CNC EDM is a special processing method, which makes use of discharge phenomenon with two different polarities of the electrode in the insulating liquid(绝缘液体), and can remove the material and then complete the processing. It has special advantages for the complex-shape molds and difficult-to-machine materials as shown in Figure 1-8.

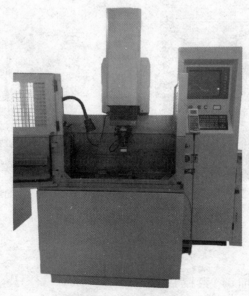

Figure 1-8 CNC EDM

7. CNC wire-cut electrical discharge machines(WEDM)(线切割机床)

The working principle of CNC WEDM is similar to CNC EDM's, the only difference is that the wire electrode replaces the electrode.

1.5 The development of machinery manufacturing systems

With the continuous improvement of automation of CNC machine tools, the automated manufacturing systems based on CNC machine tools have become the key in the industrialized countries.

1. DNC system

The earliest definition of DNC referred to direct NC, which was researched in the 1960s. At that time, NC systems were expensive and punch tapes were easy to be broken, so the DNC systems were just some NC devices directly connected to a central computer, which was used to manage and transfer NC programs.

Since the 1970s, with the continuous development of CNC technology, the storage capacity and calculation speed of CNC systems had been greatly improved. The meaning of DNC was reformed from the simple direct NC to distributed NC. It not only includes all the functions of direct NC, but also has other advanced functions, such as the system information collection, system status monitoring, system control, and so on.

Since the 1980s, with the rapid development of the computer and communication technology, the meaning and functions of DNC had been widely expanding. Compared to the DNC system in the 1960s and 1970s, the new generation of DNC systems began to focus on the information integration of the workshop, aiming at the production plans, technical preparation, processing operations and other basic operations, hence realized the centralized monitoring and distributed control. The production tasks were dispatched to the processing units through the local area network, and the information was exchanged between them. The material handling and other systems could be extended condition permitting, thus it was not only suitable for the existing production environment to improve productivity, but also saved costs.

The main components of a DNC system are showed in Figure 1-9, including central computer and peripheral storage devices, communication interfaces, machine tools and machine controllers. The central computer is employed for data management, which transfers the machining programs from the large-capacity memory to the machine tools. The bidirectional information flow is controlled to dispatch data among multiple computers in order to achieve individual processing of each machine controllers. Finally, the central computer monitors and handles the feedback information from the machine tools. The information exchange and interconnections are the key issue for the DNC system. The main difference between DNC and FMS (flexible manufacturing system) is that there is no automated material handling system in the DNC, which makes it low cost and easy to implement.

DNC is suitable for multiple CNC machine tools, usually 4—6 or more and the manufacturing environment with NC program management problems (NC programs are too large for the CNC to storage, or the CNC requires changing programs frequently in machining, etc.).

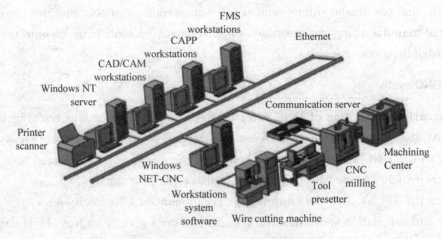

Figure 1-9　Workshop DNC model

2. Flexible manufacturing system (FMS)

According to the military standard of China, "the FMS is an automated manufacturing system which includes CNC machining equipment, material transportation devices and computer control system. It includes several flexible manufacturing cells (FMC) to achieve rapid reconfigure according to the change of manufacturing tasks or the production environment, which is suitable for multi-type, small and medium-volume production." In short, FMS is an automated manufacturing system composed of a number of CNC devices, material handling devices and computer control systems, and can be rapidly adjusted according to the manufacturing tasks and changes in the production type. At present, the most commonly seen FMS configuration includes four or more automatic CNC machine tools (machining centers and turning centers, etc.) connected by the centralized information control system and material handling system, which can achieve multi-type, small and medium batch processing and management without stopping.

FMC is the smallest scale FMS, which is the result of the development of FMS towards low-cost and small size. It has one or a small number of machining centers, industrial robots, CNC machine tools and material handling devices. The FMC has independent automatic processing functions, and also some automatic transport, monitoring and management functions, which can achieve some specific multi-type small-batch processing. Some FMCs have also achieved 24 hours unmanned running. It is more suitable for the small and medium companies with limited financial resources. Nowadays, many of the manufacturers have focused on the development of FMCs.

A flexible manufacturing factory (FMF) integrates a number of FMSs together with an automated multi-layer warehouse, and uses a computer system for communication, which contains the ordering, design, processing, assembling, inspection and delivery functions to form a complete FMS. It includes CAD/CAM, and makes the computer integrated manufacturing system (CIMS) into reality, to achieve the flexibility and automation for the production system, and then to carry out the production management, product processing and material handling in the factory scope. FMF is the highest level of automated production, reflecting the most advanced automation technology in the world, and also provides the basis for the realization of CIMS.

In order to ensure the continuous operation of the FMS, the cutting tools and the cutting process must be monitored. The possible methods include metering the output current, power or torque of the spindle motor; monitoring the signals of the tool breaking with sensors; directly measuring the changes in the cutting tool blade or workpiece size with a probe; accumulating the cutting time of the cutting tool for tool life management. In addition, the contact probe can be used to measure the thermal deformation of the machine tools and the installation error of the workpiece, and compensate the error accordingly.

3. Computer-integrated manufacturing system (CIMS) (计算机集成制造系统)

In 1974, Dr. Joseph Harrington firstly pointed out the concept of computer-integrated manufacturing (CIM) in the book titled "Computer-Integrated Manufacturing": a system based on the computer integrated manufacturing is called a CIMS. CIM is a philosophy or a guiding principle for organizing modern productions, and CIMS is the realization of this philosophy. The core content of CIM is, using the computer hardware, network and database technology, to integrate the business operation, management, planning, product design, manufacturing, sales and service departments, and the people, financial and material, in order to achieve high efficiency, high quality and high flexibility of the management, hence improve the competitiveness of enterprises. It focuses on solving the problems of system information integration in product design and business management, which combines the information technology, management technology and manufacturing technology for shortening product design, development, and manufacturing cycle, and achieving better adaption to the diversified market demands.

The entire production process is essentially about data collection, transfer and processing. The final product can be seen as the material representation of the computer data.

CIMS consists of four functional subsystems:

(1) Management information subsystem (MIS): Support the production planning and control, sales, purchasing, warehousing, accounting and other functions, and

dealing with the information of production tasks.

(2) Technical information subsystem (TIS): Product design and manufacturing engineering design automation or computer-aided design (CAD) system, computer-aided process planning (CAPP) and other subsystems to support product design and process preparation and other functions, dealing with the information on product structure.

(3) Manufacturing automation subsystem (MAS): Manufacturing systems of different automation levels, such as NC machines, FMS, and other manufacturing units, which are used to achieve the control of material flow by information and conversion of material. They are the combination of information flow and material flow used to support the manufacturing function of the enterprise.

(4) Computer-aided quality (CAQ) assurance subsystem: Used to support the quality management and quality assurance functions of the production process, it not only deal with the management information (such as rejection rate), but also process the technical information (such as measuring the performance of products).

A CIMS just utilizes computer network and database subsystem to integrate the information to achieve modern and efficient manufacturing.

The functions of the CIMS shown in Figure 1-10, include:

(1) Management functions: Enable scientific business decision for the enterprise;

(2) Engineering design automation: Uses CAD/CAPP/CAM to improve the product design and production capacity;

(3) Manufacturing automation: Uses FMC, FMS and other advanced technology to improve the quality and flexibility of manufacturing;

(4) Quality assurance: Manages and ensures the quality of the manufacturing process, reduces rejection rate and improves product performance.

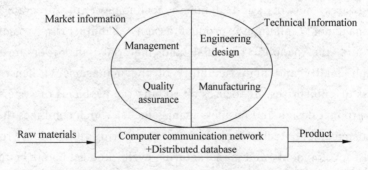

Figure 1-10 Functional diagram of CIMS

Chapter 2 Computer numerical control (CNC) systems

2.1 The working principle of the CNC system

The core of CNC machine tool is the CNC system (数控系统 shù kòng xì tǒng), which is a computer control system based on special software. From the automatic control, CNC system is a trajectory control system i.e. tool follows a pre-defined path. In essential, work can be performed in all axes simultaneously with the help of multi-actuators.

2.1.1 Development of the NC system

1. The history of CNC system

Since 1950s, the NC machine tool has developed for more than 60 years. Current CNC system is evolved as the result of 6 generations development comprised of two major stages, as shown in Table 2-1.

Table 2-1 Evolution of NC system

Classfication	Generation	The date of birth	System components
Hardware NC	First generation	1952	Electron Tube
	Second generation	1959	Transistor
	Third generation	1965	Small-scale integrated circuit
Computer numerical control (CNC)	Fourth generation	1970	Small computer
	Fifth generation	1974	Microprocessor
	Sixth generation	1990	Generation based on personal PC

1) Numerical control (NC) phase (1952—1970)

Early, the computer operation speed is slow, which cannot meet the requirements of real-time control of machine tools. Digital logic circuit had to be taken into a machine tool as a NC system, which is known as the hard-wired (硬连线的 yìng lián xiàn de) NC. With the development of electronic components, this stage has gone through three generations, including the electron tube generation, the transistor generation and the small-scale integrated circuit generation.

(1) The electron tube(电子管)(diàn zǐ guǎn) is the vacuum tube, see Figure 2-1. No matter what vacuum tube is diode, triode or more electrodes, its structure is the same, including vacuum glass shell (metal or ceramic), filament, the cathode and the anode. To the electronic tube radio, for example, only five or six electronic tubes are used and the output power is only about 1 W against input of 40—50 W. Turn on the power switch and wait for more than 1 min before it rings. At that time, power consumption and control speed were almost unimaginable.

Figure 2-1　Vacuum tube

(2) Transistors are important components to control the current in a circuit. See Figure 2-2, the transistor was invented in the Bell laboratory. The invention of the transistor has important significance for the technology revolution and innovation. Compared with the vacuum tube, the transistor is much cheaper, more durable(耐用的)(nài yòng de) and lower energy consumption, while it can be reduced in a smaller size. The transistor is the unit of integrated circuit and chip.

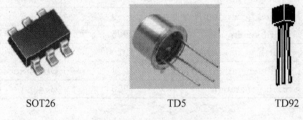

Figure 2-2　Transistors

(3) Small-scale integrated circuits. It is formed by etching micro transistors on silicon wafers. Because of its small size and low power consumption, the reliability of CNC system is further improved.

2) Computer numerical control (CNC) stage

CNC stage has also gone through three generations. The fourth generation is based on minicomputer, the fifth on microprocessor and the sixth on PC (PC-based).

(1) In 1970, the general-purpose minicomputer had appeared and could be batch produced. So it was transplanted as a core component of the CNC system and the CNC stage began. In 1971, the microprocessor was created by the United States INTEL company. It was the first time that the two most core computer components, i.e. computing and controller, are integrated on one chip with large-scale integrated circuit technology, also known as the central processing unit (CPU).

(2) In 1974, the microprocessor was applied to the NC system. Because the function of minicomputers was so strong that it is wasteful to only control one machine tool. So at that time, it was used to control more than one machine tool, known as group control. It is more economic to use the microprocessor. And the reliability of the minicomputer was not better. Although the speed and function of the early microprocessor was not good enough, it could be solved by the multiprocessor architecture(多处理器体系结构).

(3) In 1990, the performance of the personal computer (PC) became better, which can meet the requirements of the core components(核心部件) of the CNC system. The CNC system based on the PC began. The common open-architecture CNC system includes CNC embedded PC, PC embedded CNC and software CNC.

It is an important milestone that the computer was used for the CNC machine tool in the history, because it is the combination of achievements in modern computer technology, automatic control technology, sensor technology and measuring(测量) technology, machinery manufacturing technology, as a result, the mechanical processing technology has reached a quite new level.

Compared with NC, CNC has many advantages. Many functions of CNC are realized by software, which can be easily reprogrammed to meet the different demands of controlled mechanical equipment, so as to realize the CNC function changes or extensions. It is convenient to the users(用户).

2. The development trend of CNC system

With the development of electronic and computer technology, the performance of CNC system is getting better and better. It is developing towards high speed, high precision, high reliability, intelligence(智能), networking, opening and so on.

1) High speed, high precision

In CNC system, speed and accuracy are two important factors, directly related to the processing efficiency and product quality(质量).

The most effective way to improve the speed of CNC is to increase the number of bits and the main frequency of the microprocessor(微处理器). For example, the CPU bits is raised from 16 to 32, now even 64 bits. The main frequency is from 5 MHz to 2 GHz. Multi-CPU system can improve the control speed. Increasing spindle speed is also an important means to achieve high speed, such as the spindle speed is up to 150,000 r/min.

High accuracy can be achieved by reducing the error of CNC system and using the compensation technique. The general method is to improve the resolution. With regard to the compensation technology, the gap compensation, screw pitch compensation and tool compensation technology are used widely. Except for those, recently hot deformation compensation had been quite concerned deeply. In order to reduce the error caused by the heat distortion of the motor for the main shaft and the screw pair(螺旋副), the flowing oil is used to cool the inner spindle motor and the bearing, while hot compensation is applied, too.

2) High reliability

With the rapid development of the network application for CNC machine tools, the high reliability of CNC system has become the primary goal. For the unmanned factory, if the MTBF (mean time between failures) of CNC machine tool is 3,000 hours, it is assumed that failure rate between a CNC machine tool and a CNC system is 10 : 1. So MTBF of the CNC system will be greater than 30,000 hours, while one of the CNC devices, such as spindle and driver's MTBF must be greater than 100 thousand hours.

3) Open architecture

For a new generation of open architecture CNC system, all the hardware, software and bus specification are open. CNC system manufacturers and users can integrate system and configure based on these resources and actual needs. Open architecture is very convenient and promotes the development and application of CNC system toward multilevel and multispecies, greatly shortening the production cycle. At the same time, the CNC system can be upgraded(升级) with the CPU, while the structure can remain unchanged. In recent years, many countries have developed this system, such as the United States Scientific Manufacturing Center and air force joint leadership of the next-generation workstations/machine controller architecture of NGC, the automation system of open architecture OSACA, Japan OSEC plan, etc. Especially, the advent of STEP-NC standards, more open architecture provides a broader space for development. A large number of general-purpose(通用的) micro computer tools can be used, so that it become both easy and quick to program, operate and upgrade.

4) Software CNC

Nowadays, the CNC systems are classified into closed architecture, PC-embedded NCs, NC-embedded PCs and soft ones. There is a growing trend towards the soft CNC systems.

Traditional CNC systems, such as FANUC0 system, MITSUBISHI M50 system, SINUMERIK 810M/T/G system, etc, are special closed architecture ones. At present, this system type still occupies the most market shares of the manufacturing(制造). However, the closed architecture makes it difficult to add sensors for process monitoring or machine control. To overcome such drawbacks of the closed CNC system, an open CNC system has been developed.

"PC-embedded NC" structure, such as FANUC18i, 16i system, SINUMERIK 840D system, Num1060 system, AB 9/360 NC system, is the achievement that a number of CNC system manufacturers combine the extensive experience of NC technology and the rich computer software resources to develop and accomplish. In some way, this structure is open(开放的), but for its NC part remains the traditional CNC system, the user can not intervene in the core of CNC system. This kind of system is complex, powerful and expensive.

"NC-embedded PC" structure consists of an open architecture motion control card and a PC. This kind of motion control card, often taking the high-speed DSP as CPU, has very strong control ability(能力) for movement and programmable logic controller (PLC). It is itself a CNC system that can be used alone. Function library is open for the user to control the system under the Windows platform. Therefore, the open architecture motion control card is widely used in various fields of automation control in manufacturing industry. For example, PMAC from the United States Delta Tau is a multi-axis motion control card (shown in Figure 2-3).

Figure 2-3 PMAC Motion control card (PC104)

The SOFT open CNC system is a new open architecture CNC system. It provides the user with the greatest choice and flexibility, and all of the CNC software is installed in the computer, and the hardware part is only a standardized(标准化的) interface to the computer, the servo driver and external input/output (I/O). It stays the same that the computer can install a variety of brand sound cards and the corresponding driver software(驱动软件). Various types CNC systems with high performance can be developed based on the open CNC kernel in the Windows NT platform and the system functions can be ordered. Compared with the previous(以前的) CNC system, SOFT open CNC one has the highest performance price ratio and the most vitality. It is becoming an important trend to replace the complex hardware by software intelligence. Its typical products are the United States MDSI Open CNC, Germany Power Automation company PA8000 NT, etc.

5) Intelligence

In 21st century, with the development of <u>artificial intelligence</u>(AI) and the penetration into the computer field, adaptive control, fuzzy systems and neural networks are introduced into the CNC system, which has automatic programming, feed forward control, fuzzy control, learning control, adaptive control, <u>process parameters</u>(工艺参数) automatic generation, 3D cutter compensation, dynamic parameter compensation and other functions, in addition, the machine interface is very friendly and has a fault-diagnosis expert system which makes the self-diagnosis and fault-monitoring function more perfect. The self-adaptive servo system can identify and optimize automatically the speed by the load.

6) Network

The network control system means that the control system connects with or controls other control systems or the host computer by the network. At first, CNC system connects with the production site by the Intranet, then with the outside of the enterprise through the Internet, which is called Internet/Intranet technology.

With the development and maturity of network technology, digital manufacturing was put forward in the industry. Digital manufacturing, also known as "e-manufacturing", is one of the symbols of the manufacturing enterprises modernization and also the international advanced machine tool manufacturers' standard configuration to supply. With the extensive use of information technology, more and more domestic users require <u>remote communication services</u>(远程通信服务).

The network CNC system promotes highly the development of flexible manufacturing technology. Modern <u>flexible manufacturing system</u>(FMS) develops from point (the single machine, machining center or CNC machine tool) and line (FMC, FMS, FTL or FML) to section workshop independent manufacturing island, CIMS and distributed network integrated manufacturing system. Now CNC machine tools even can be easily connected with CAD, CAM, CAPP, MTS by the networking.

7) Multi-function

In the process of machining, the considerable time is consumed in the workpiece handling, loading and unloading, installation and adjustment, changing the knife and <u>the ascending</u>(提升) and descending of spindle. In order to reduce this time as much as possible, one kind of machine tools was developed integrating different sorts of processing in the same machine. Therefore, the composite function of machine tools is the type of development in recent years.

Composite machining refers to different processes finished automatically in one clamping.

Such processes include milling, drilling, boring, grinding, tapping, and reaming(铰削), etc.

At present, the mainstream(主流) CNC system developers can provide high-performance composite CNC system.

3. Introduction to typical NC system

The world's major CNC systems include FANUC (Japan), SIEMENS (Germany), A-B (USA), FAGOR (Spain), HEIDENHAIN (Heidenhain), MITSUBISHI (Japan), NUM (Switzerland), etc. In China, there are Guangzhou CNC system, KND CNC, Shenyang CNC, DASEN, etc.

SIEMENS CNC systems will be introduced, including SINUMERIK3/8/810/820/850/880/805/840 series. SINUMERIK 802S, 802C, 802D, 810D, 840D are commonly used, whose instructions can be downloaded from the site http://www.ad.siemens.com.cn/products/. Appearances are as shown in Figures 2-4 and 2-5. These CNC systems belong to the same family and have certain inheritance(继承).

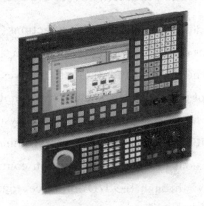

Figure 2-4　SINUMERIK 802S　　　　Figure 2-5　SINUMERIK 840D

SINUMERIK 840D is a high-performance CNC system at the end of the 1990s launch of the SIEMENS company, it maintains three CPU structures: man-machine communication CPU (MMC-CPU); digital(数字的) control CPU (NC-CPU); programmable logic controller CPU (PLC-CPU). The three parts functionally divide and support each other.

2.1.2　The composition and working principle of the CNC system

According to the definition from NC Standardization Committee of U.S. Electronics Industry Association (EIA), the CNC system is a special computer system through the implementation of the program in the memory to complete the part or all functions for the requirements of NC, and is equipped with a servo drive interface circuit. CNC system can change the control(控制) function only through changing the software but not changing the hardware. Thus, the CNC system is used widely, showing greater flexibility.

From the definition, the CNC system is composed of a program, an I/O device, a CNC device, a programmable logic controller (PLC), a spindle drive and a servo drive device. The I/O device, also known as HMI, includes a tape reader (outdated), a paper tape punch (rare), a keyboard, a display operation panel, and external storage devices, etc.

The core component of the NC system is the CNC device. This device is composed of hardware and software(软件). The hardware consists of CPU, memory, position control, input/output interface, PLC (built-in), graphic control, power supply module; while the software refers to management and control programs. Compared with the NC device, the CNC device has more powerful functions, and is more suitable for the complex control requirements of CNC machine tools.

The main function of PLC is to receive the instructions of the auxiliary function or the operation instruction from the operation control panel, controlling the interlock(联锁) and all kinds of auxiliary actions(辅助措施), as well as displaying the states of various control signals.

The spindle and the servo drives are composed of position control unit, speed control unit and feed/spindle servo motor.

The working principle of CNC system is shown in Figure 2-6. CNC system receives the NC program, then decodes, undergoes data processing, and interpolates; then carries out instructions to position control, spindle control and auxiliary functions, and finally, through the I/O interface, outputs, controls, displays and diagnoses, step by step. The display or indicator shows corresponding parameters, such as current location, spindle speed, coolant state and so on. The diagnostic procedure can be carried out online, off-line or remotely(远程地).

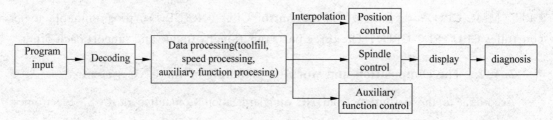

Figure 2-6 Schematic diagram of the CNC system

Decoding is to translate the CNC program into the code that can be identified by the NC unit.

Data processing generally includes tool radius, length compensation, speed calculation and auxiliary function processing.

The tool radius compensation is to transform the contour of the part into the tool center track. The tool length compensation is to convert Z-axis number from program into Z-direction number of a fixed point from the tool holder. Because the CNC system only controls the tool center and fixed point from tool holder which is set before the CNC machine comes out of the factory, compensation function is necessary. And then the programmer only needs to program the part contour to reduce the workload(工作量).

Velocity calculation(计算) is to solve the speed of the machining program. The feed speed of the program is the synthetic speed(合成速度), so the speed parts of every axis must be calculated according to the coordinates of every movement direction. In addition, the minimum and the maximum speed limits of the machine tool must be judged and processed.

Auxiliary functions, such as the tool changing, the spindle start/stop, cutting fluid switch, are processed in this part. The main task of the auxiliary functions is to identify the signs and send signals to the corresponding parts of the machine tool.

2.1.3 The hardware of the CNC unit

The hardware of CNC devices is generally divided into the single processor and the multi-microprocessor. Early CNC and nowadays some economical CNC systems use a single microprocessor structure. But now, the multi-microprocessor structure is developed rapidly.

1. Single-microprocessor architecture

In the single-microprocessor architecture, usually, there is only one microprocessor used for centralized control. Even, there are more than one microprocessors in some CNC system, but only one microprocessor can control the system bus and the other are the special intelligent components which cannot access the main memory. This structure also belongs to the single-microprocessor structure.

A single-microprocessor architecture is shown in Figure 2-7. It is mainly composed of a central processing unit (CPU), the memory, the bus, peripherals, and the I/O interface circuit. This system is basically the same with the personal computer. But for CNC system, output devices are different, which can convert and amplify(放大) each axis data to drive the machine table or tool (load), so as to realize the cutting motion.

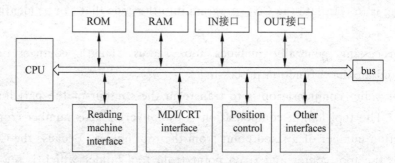

Figure 2-7　Microprocessor CNC system block diagram

2. Multi-microprocessor structure

The structure of a multi-microprocessor is composed of two or more microprocessors. Each microprocessor shares the memory, I/O interface or system bus. All microprocessors will complete the task simultaneously, so the processing speed of the whole system is greatly improved.

The CNC unit of multi-microprocessor mostly adopts modular structure. The microprocessor, memory, and I/O control are made of board (called hardware module), while the corresponding software is based on the modular structure. The CNC device includes lots of special-purpose modules used for management, interpolation, position control, PLC, operating and controlling data, I/O, display and memory respectively. The multi-microprocessor structure can be classified as the memory-shared (Figure 2-8) and the cable-shared (Figure 2-9) according to the communication among modules with each other.

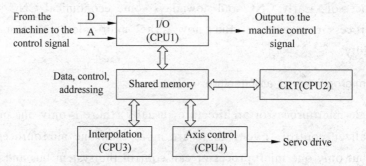

Figure 2-8　Multiprocessor-shared memory structure

The shared bus architecture is flexible, simple and easy for implementation. Its disadvantage is that the main modules will cause "competition" so as to reduce the efficiency of the transmission.

2.1.4　Hardware composition of CNC devices

The CNC device is composed of CPU, BUS, memory, I/O interface and so on (same to a microcomputer).

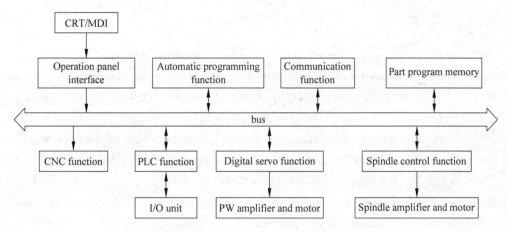

Figure 2-9 Multiprocessor-shared bus architecture

1. Central processing unit (CPU)

CPU is the core and the "brain" of the CNC system. The arithmetic/logic operations(算术/逻辑运算) are main functions of CPU. It can save a small amount of data. It can translate the code and execute the instruction, and also exchange data with I/O devices and memory. It provides timing and responses the pulse requests sent from other parts.

In the CNC units, CPU data width may be 8, 16, 32 or 64 bits. CPU meets the real-time requirements of software implementation, which is mainly reflected in the CPU word length, computing speed, addressing capability (寻址能力), interrupt service, etc.

2. BUS

The bus is a common channel for transmitting data or exchanging information. The standard bus is applied to connect CPU with other parts such as memory board, I/O interface board and so on. Various bus protocols include Profi-Bus, CAN Bus, and InterBus-S.

3. ROM (read-only memory)/RAM (random access memory)

CNC system software, part programs, raw data (原始数据), parameters, intermediate results, and processed results are stored in the ROM/RAM.

ROM can store system software. The information in it cannot be overwritten when the system is running. But RAM stores the information that may be written and overwritten.

4. I/O interface

In industries, there are two types of control methods. One is "path control" to control the feed and position of every axis completed commonly by the CNC unit, another is the "order control" finished by PLC or NC unit. "Order control" is to control machine stroke switch, sensor, relay switch button, etc. according to the predetermined logical orders, such as spindle starting/stopping, reversing, tool replacing, workpiece clamping/loosening, hydraulic controlling, lubrication cooling, and so on.

The signals of the controlled equipment can be divided into switch, analog and digital ones. These signals cannot be directly associated with the CNC unit, so I/O interface circuits are needed, whose purposes are as follows:

1) The corresponding conversion or power amplification of the signal

The input signal must be changed into numerical one to meet the computer input signal requirement. While the output one should meet the various elements to control outer devices. The signal conversion mainly includes level conversion, digital-to-analog conversion(数模转换), digital and pulse conversion and power matching(功率匹配).

2) Blocking external interference signals into the computer

The CNC device is electro-optically isolated from the external signal to improve the reliability of the CNC device.

2.1.5 Software components of the CNC system

Hardware is the foundation and software is the soul. The CNC system software consists of management software and control software, as shown in Figure 2-10.

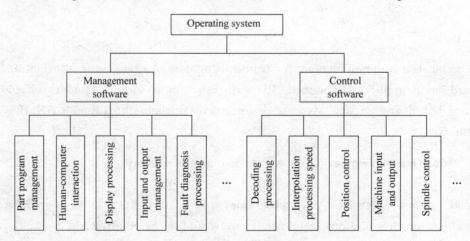

Figure 2-10 CNC device software function diagram

Management software is to manage input, I/O processing, communication, display, diagnosis and program management(程序管理).

Control software is to finish the decoding, tool compensation, speed processing, interpolation and position control.

For the single-microprocessor CNC system, front/back platforms and interrupt type software architecture are used.

For the multi-microprocessor CNC system, the operating system is fully used.

CNC software is a typical and complex real-time control system. Many of control tasks, such as the program input and decoding, tool radius compensation, interpolation, position control and precision compensation are achieved by the software. Logically, these tasks can be regarded as different functional modules, and there are many coupling relationships between modules. In the design of CNC software, it is necessary to consider how to organize and coordinate these functional modules to meet certain timing and logical relationships.

Real-time operating system is an important branch of the operating systems, it has not only the function of the general operating system, but also the functions of task(任务) management, real-time scheduling mechanism, communication mechanism between tasks and so on.

At present, there are two kinds of models. First is real-time operating system, such as RTOS, /COS-II system. Second is the general PC operating system (DOS, Windows, LINUX, or UNIX) -embedded real-time operating system. Some domestic manufacturers and scientific research institutions adopt the latter.

2.1.6 The functions of CNC devices

The functions of CNC system include the basic and the additional ones.

The basic functions of the CNC system configuration are the necessary ones, dedicating to the control, the preparation, the interpolation, the feed, the spindle, the cutting tool management, the auxiliary task (M instruction), or the character display.

The additional features dedicate to compensation, cycle fixing, graphic display (图形显示), communication, as well as man-machine dialogue programming.

The following is to explain the above functions in detail.

1. Control function

The CNC can control moving shafts and rotary shafts. For example, the CNC lathe needs at least two-axis linkage(联动), and the lathe with the multiple tool holder needs more than one axis to control. For the CNC boring milling machine and machining center, three or more shafts are needed to be controlled. As the number of linkage control axes increases, the CNC system will be more complex and its programming will be increasingly difficult as well.

2. Preparation function(准备功能)

That is G function which includes instructions of the machine tools action mode.

3. Interpolation functions and fixed-cycle function

The interpolation function is to realize the machining of the outline of the part (plane or space). General CNC system only has linear and circular interpolation, and now the more high-end(高端的) CNC system is also equipped with parabola/ellipse/polar coordinates, sine/spiral lines and spline interpolation functions.

In the machining, some processes, such as drilling, thread cutting, tapping(攻螺纹) and boring, require to complete the action cycles which are very typical and repeated many times, so the CNC system defines these typical fixed-cycle codes in advance which can be directly used to complete these machining. The fixed-cycle function can greatly simplify the programming work.

4. Feed function

The feed-speed control function mainly has the following three kinds:

(1) Feed rate: control the speed of the tool relative to the workpiece, unit mm/min.

(2) Synchronous feed speed: control the synchronization between the spindle speed and the feed speed, used for processing thread, unit mm/R.

(3) Adjustable feed rate: Feed rate can be adjusted manually in real-time. That is, the pre-set(预设置) feed rate can be changed by turning the rate band switch (0%—200%) when machining.

5. Spindle control function

Spindle control includes the following:

(1) Spindle speed: to control the cutting point of tool, unit r/min.

(2) Constant line speed: to control the spindle, a constant used to turn the end face, unit m/min.

(3) The C-axis control: The spindle can stop at any position among the circumferential direction.

(4) Spindle adjustment rate: manual real-time adjustment. That is, the pre-set spindle speed can be changed by turning the rate band switch (0%—200%) when machining to get the real-time spindle speed.

6. Auxiliary function

The auxiliary functions are M codes used for auxiliary(辅助的 fǔ zhù de)operation control.

7. Tool management function

To manage tool geometry dimensions and tool life, the machining center should have this function. The cutting tool geometry refers to tool radius and length, which can be used to compensate. Tool life generally refers to the time of life. When a tool life expires, the CNC system will promote the user to replace the tool. In addition, the T codes are used to manage tools, marking and selecting the tools automatically.

8. Compensation function

(1) Tool radius and length compensation: With this function, the tool radius, length or wear or replacement can be compensated(补偿 bǔ cháng). This function is implemented by the G command.

(2) Transmission chain error: The pitch error and backlash error can be compensated. Input the prior measurement of pitch error and backlash into the corresponding storage unit and then compensate.

(3) Intelligent compensation: Such as the processing errors, caused by machine geometry, and thermal/elastic deformation, and machining error caused by tool wear; all of them can be online compensated by the model developed by means of the modern advanced AI and expert system. This belongs to the researching and developing technology.

9. Man-machine conversation function

In the CNC device, monochrome/color CRT is equipped to display the characters and graphics(图形 tú xíng), facilitating the users' operation and use. This kind of function includes the operating interface of menu, the editing environment of parts processing program, system and displaying, querying or modifying to the machine parameters, status, fault information, etc.

10. Self-diagnosis function

Generally, the CNC device has the function of self-diagnosis, which can be realized by software especially for the modern CNC device. This function can quickly identify the types and locations of the faults, so as to facilitate timely troubleshooting and reduce downtime.

Diagnostic procedures can be included in the program checking during the operation of the system, and also used as a service program diagnosing before the system operation or after fault. Some CNC devices have the remote communication diagnosis(诊断 zhěn duàn) system.

11. Communication function

This function is for the CNC device to exchange information with the outside world. Usually the CNC device can communicate with the superior computer with RS232C to transmit parts processing programs. Some devices also have DNC interface to realize the direct NC. The more advanced system can relate with MAP (manufacturing automation protocol) which adapts to the integration with the other large manufacturing systems, such as FMS, CIMS, IMS and so on.

2.2 Interpolation theory

2.2.1 Overview of the interpolation

1. Interpolation definition(插补定义 chā bǔ dìng yì)

During the machining, generally, users only provide the necessary relevant parameters(参数 cān shù)describing the profile, such as the starting/end coordinates for a line, while starting/end coordinates clockwise and circle-center coordinates or radius for arc. But these parameters cannot meet the control demands of the operating components (such as stepper motor, AC and DC servo motors). Therefore, in the contour control system, a number of intermediate points must be calculated in real time according to the profiles (G01, G02) and the feed rate (F). This is the concept of interpolation. It means that the microline segments are calculated and the tool will move along the microline. After a number of interpolation cycles, the tool moves from the start point to the end point and then the complete contour is machined. Interpolation can be implemented by different calculation methods called interpolation algorithms. The essence(本质 běn zhì) of interpolation is the densification of data points.

The interpolation is the most important function and essential feature in the contour control system. The stability and accuracy of the interpolation algorithm will directly affect the performance of CNC system. So, in order to make the best performance, scientific research personnel always develop the new interpolation algorithm (software) with higher precision and speed which is the well-kept secret for the CNC system company.

2. Interpolation classification

The interpolators can be classified as either hardware or software ones. Hardware interpolation, as long as the parameters and interpolation commands are given out, the whole process can be automatically controlled by the microprocessor without any interference from the software. The high-speed microprocessor and field-programmable gate array (FPGA) are currently used to implement hardware interpolators for high-speed um-level control, such as the Guangzhou GSK980TA lathe CNC system. Software interpolation runs slightly slower, but its structure is simple and flexible. Modern NC system mostly adopts software interpolator. In some cases, hardware interpolation can also be used as the second level of interpolation when software and hardware are combined with the interpolation. But whether it is software or hardware ones, its arithmetic principle(原理 yuán lǐ) is basically the same. The straight line and arc are the basic components of the contour, so most CNC systems have linear and arc interpolations. In some advanced CNC system, parabolic, spline, spherical or helical interpolations(螺旋插补 luó xuán chā bǔ) have been included. There are many interpolation methods according to the different principles and calculation methods. At present, the interpolation methods can be classified into the reference-pulse and the sampled-data interpolation methods.

1) Reference-pulse interpolator

Reference-pulse interpolation is also called stroke-scalar interpolation or pulse-incremental interpolation. During the interpolating, the CNC device will output reference pulse sequence(序列 xù liè) to each motion coordinate. This interpolation method is relatively simple (only addition and shift) and can be implemented by hardware. With the rapid development of computer technology, software is used to complete this algorithm. The number of the pulse represents the position of the moving axis, and the frequency of the pulse is proportional to the velocity of the moving axis. Because the maximum speed of the rotating shaft is limited by the time of the interpolation algorithm, it is only suitable for the economical CNC system with the medium-speed requirements. The pulse equivalent is 0.01 mm for the ordinary-precision machine tools and more precise one reaches 1 μm or 5 μm. The feed speed is generally 1—3 m/min. This method is mainly used in the open-loop control system driven by stepper motor, and also used in the fine interpolation of sampled-data interpolation.

The reference-pulse interpolation method can be classified into digital pulse multiplier(乘数 chéng shù) interpolation method, point by point comparison method, digital integral method, compared with the vector discriminant method, integral method, the minimum deviation method, the target tracking method, direct function method, step tracking method, encryption discrimination and double discriminant interpolation method and Bresenham algorithm.

In general, the minimum deviation method has higher accuracy(精度) and it is favorable to the continuous motion of the motor.

2) Sampled-data interpolator

Data-sampling interpolation is also called time-scalar interpolation or digital-incremental interpolation. Its algorithm characteristic is that the CNC device produces not a single pulse(单脉冲), but a standard binary word per interpolation cycle. The interpolation operation is completed by two steps. The first is rough interpolation, and it is inserted into a plurality of points between the starting and end points within the given curve. A number of small line segments will be gotten to approximate the given curve, all micro-line lengths are equal, associated with a given feed rate F and the interpolation period T. The second is the fine interpolation, which is similar with the pulse-increment interpolation. In each sampling period, the actual position is sampled and compared with the calculated values of the interpolation. The speed of the corresponding shaft is calculated according to the compared error. Generally, the rough interpolation will be finished by software. The fine interpolation can be implemented in software or hardware. The sampled-data interpolation is suitable for the closed-and semi closed-loop position control systems based on DC and AC servo motors. For the data sampling interpolation, the interpolation cycle directly affects the position control accuracy and feed rate.

The data sampling interpolation includes direct function, expansion of digital integral, two-order recursive extended digital integral circular interpolation, double arc digital integral interpolation, angle approaching circular interpolation and the improved Tusiding. In recent years, many scholars have studied more types of interpolation and improved methods, such as improving DDA arc interpolation algorithm, space arc segmentation, parabolic time-division interpolation(抛物线时间分割插补法), elliptical arc interpolation method(椭圆弧插补法)and so on.

With the development of geometric modeling technology, most CAD/CAM systems use parametric curves and surfaces to represent the shape of workpiece, so parametric curves interpolation becomes the focus of research. The interpolation process of CNC system is to get continuous coordinates, and the points on the curve is corresponding with the parameter values, so the parametric curve interpolation can be transformed into solving a series of parameter values. Parametric curve interpolation includes Bezier, B spline, Non-Uniform Rational B-Spline (NURBS) and other parameters interpolation methods for arbitrary spatial parameter curves. Compared with the traditional algorithm, the real-time interpolation of parametric curve can obtain higher machining precision.

With the promulgation(颁布) of STEP standards, the NURBS curve(曲线) and surface interpolation method will be applied widely. Because the NURBS can express conic curves, free curve and surface, in the future, the CNC system interpolation can include only the straight line and NURBS.

In the study of interpolation algorithm, the stability not only must be taken into account, but the calculation speed and accuracy of interpolation.

2.2.2 Point by point comparison interpolation

With the point by point comparison interpolation algorithm (逐点比较插补算法), the points are calculated one by one by the following four steps.

1) Deviation judgement

In order to determine the direction of tool feed, it is necessary to judge the deviation of the current tool position with respect to the given contour;

2) Feed control

Based on the deviation, the tool will feed one step relative to the workpiece(工件) contour to reduce the deviation;

3) New deviation calculation

Because the tool has changed the position after feeding, the new position of the current position of the tool should be calculated to prepare for the next deviation;

4) Judgment of end point

It is to determine whether the tool has reached the end of the processed contour. If the tool has reached the end point, then stop interpolation; if not, continue.

1. Linear interpolation principle of point by point comparison method

1) Linear interpolation principle of the first quadrant

(1) Deviation judgment

Taking the straight line of the first quadrant as an example. In the program, the starting/end coordinates of the line are given. If the starting point is at the origin(原点) of coordinates, the end point are (x_e, y_e), and the interpolation points are $(x_i, y_i)(i=1,2,3)$, as shown in Figure 2-11.

Straight line OP_e, the angle between OP_e and x-axis is α_e, while the angle between any point and x-axis is α_i, then

$$\tan\alpha_e = y_e/x_e$$

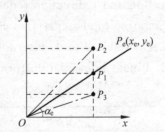

Figure 2-11 The relationship between interpolation points and straight line

$$\tan\alpha_i = y_i/x_i$$

If the interpolation point $P_1(x_i, y_i)$ is just in a straight line, then
$$\tan\alpha_e = \tan\alpha_i$$
$$f_i = y_i x_e - x_i y_e = 0$$

If the interpolation point $P_2(x_i, y_i)$ is above the line, then
$$\tan\alpha_i > \tan\alpha_e$$
$$f_i = y_i x_e - x_i y_e > 0$$

If the interpolation point $P_3(x_i, y_i)$ is below the line, then
$$\tan\alpha_i < \tan\alpha_e$$
$$f_i = y_i x_e - x_i y_e < 0$$

To sum up, the deviation function $f_i = y_i x_e - x_i y_e$.

If $f_i = 0$, then the interpolation point (x_i, y_i) is just in line;

If $f_i > 0$, then the interpolation point (x_i, y_i) is above the line;

If $f_i < 0$, then the interpolation point (x_i, y_i) is below the line.

(2) Feed control

When $f_i \geq 0$, the step moves in the direction of $+x$;

When $f_i < 0$, the step moves in the direction of $+y$.

(3) New deviation calculation

The multiplication operation inside the computer is more time-consuming than the addition, so the computation of the discriminant function f is realized by the recursive superposition method.

When one step is added along $+x$, then
$$f_{i+1} = y_{i+1} x_e - x_{i+1} y_e = y_i x_e - (x_i + 1) y_e = f_i - y_e$$

Similarly, when one step is added along $+y$, then
$$f_{i+1} = y_{i+1} x_e - x_{i+1} y_e = (y_i + 1) x_e - x_i y_e = f_i + x_e$$

(4) End-point judgement

One-way judgement: take the longer coordinate as the count length;

Two-way judgement: take every coordinate as count lengths, until the sums of both x direction and y direction reduce to 0, the interpolation will stop.

Counting total steps: that is to get x and y total number of steps (denoted A). Until A is reduced to 0, interpolation will stop.

2) Linear interpolation formula and graph of each quadrant is shown in Table 2-2.

Table 2-2 Linear interpolation formula for each quadrant

	$f_i \geq 0 (i=0, 1, \cdots)$	$f_i < 0 (i=1, 2, \cdots)$	Graphical
First Quadrant	$(+\Delta x)$ $f_{i+1} = f_i - (y_e - y_0)$ $x_{i+1} = x_i + 1,\ y_{i+1} = y_i$	$(+\Delta y)$ $f_{i+1} = f_i + (x_e - x_0)$ $x_{i+1} = x_i,\ y_{i+1} = y_i + 1$	

	$f_i \geqslant 0 (i=0, 1, \cdots)$	$f_i < 0 (i=1, 2, \cdots)$	continuous Graphical
Second Quadrant	$(-\Delta x)$ $f_{i+1} = f_i - (y_e - y_0)$ $x_{i+1} = x_i - 1, \ y_{i+1} = y_i$	$(+\Delta y)$ $f_{i+1} = f_i + \|x_e - x_0\|$ $x_{i+1} = x_i, \ y_{i+1} = y_i + 1$	
Third Quadrant	$(-\Delta x)$ $f_{i+1} = f_i - \|y_e - y_0\|$ $x_{i+1} = x_i - 1, \ y_{i+1} = y_i$	$(-\Delta y)$ $f_{i+1} = f_i + \|x_e - x_0\|$ $x_{i+1} = x_i, \ y_{i+1} = y_i - 1$	
Fourth Quadrant	$(+\Delta x)$ $f_{i+1} = f_i - \|y_e - y_0\|$ $x_{i+1} = x_i + 1, \ y_{i+1} = y_i$	$(-\Delta y)$ $f_{i+1} = f_i + \|x_e - x_0\|$ $x_{i+1} = x_i, \ y_{i+1} = y_i - 1$	

Linear interpolation process of the first quadrant point by point comparison method is shown in Figure 2-12.

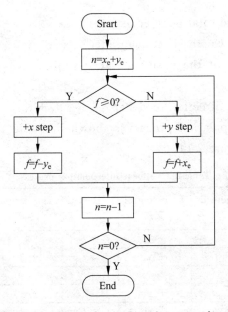

Figure 2-12 Flow chart of point by point comparison method

3) Linear interpolation not at the origin

If the start point is not at the origin, the theory is the same as the start point from the origin. The start point is (x_0, y_0), the end point is (x_e, y_e), set $P(x_i, y_i)$ as a fixed point, there is:

$$\frac{x_i - x_0}{y_i - y_0} = \frac{x_e - x_0}{y_e - y_0}$$

Thus available:

$$f_i = x_i(y_e - y_0) - y_i(x_e - x_0) - x_0(y_e - y_0) + y_0(x_e - x_0)$$

When moving toward $+\Delta x$,
$$f_{i+1} = x_{i+1}(y_e - y_0) - y_{i+1}(x_e - x_0) - x_0(y_e - y_0) + y_0(x_e - x_0)$$
$$f_{i+1} = (x_i + 1)(y_e - y_0) - y_i(x_e - x_0) - x_0(y_e - y_0) + y_0(x_e - x_0)$$

Tidy up,
$$f_{i+1} = f_i + (y_e - y_0)$$

When moving toward $+\Delta y$,
$$f_{i+1} = f_i - (x_e - x_0)$$

When $f_i = 0$, the processing point is just in the straight line. When $f_i > 0$, the point is up to the straight line. When $f_i < 0$, the processing is down to the straight line.

4) Linear feed direction

The symbols and feed directions of four quadrants are shown as Figure 2-13.

5) Case for point by point interpolation method

Example: The pulse equivalent is 1, the starting point (0, 0), the end point (5, 3).

Solution:

(1) $x_e + y_e = 8$, so the total number of steps is 8;

(2) The straight line for the first quadrant, $f_i \geq 0$, the direction of the positive x feed step, $f_{i+1} = f_i - y_e$;

$f_i < 0$, the positive y direction feed step, $f_{i+1} = f_i + y_e$;

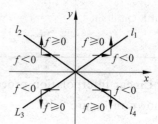

Figure 2-13 Symbols and feed directions of four quadrants

(3) The linear interpolation process is shown in Table 2-3 and the interpolation trajectory is shown in Figure 2-14.

Table 2-3 Linear interpolation process

Serial number	Deviation judgement	Feed control	Deviation calculation	End point discrimination
1	$f_0 = 0$	$+\Delta x$	$f_1 = f_0 - y_e = 0 - 3 = -3$	$M = 8 - 1 = 7$
2	$f_1 < 0$	$+\Delta y$	$f_2 = f_1 + x_e = -3 + 5 = 2$	6
3	$f_2 > 0$	$+\Delta x$	$f_3 = f_2 - y_e = 2 - 3 = -1$	5
4	$f_3 < 0$	$+\Delta y$	$f_4 = f_3 + x_e = -1 + 5 = 4$	4
5	$f_4 > 0$	$+\Delta x$	$f_5 = f_4 - y_e = 4 - 3 = 1$	3
6	$f_5 > 0$	$+\Delta x$	$f_6 = f_5 - y_e = 1 - 3 = -2$	2
7	$f_6 < 0$	$+\Delta y$	$f_7 = f_6 + x_e = -2 + 5 = 3$	1
8	$f_7 > 0$	$+\Delta x$	$f_8 = f_7 - y_e = 3 - 3 = 0$	0

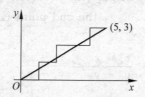

Figure 2-14 Interpolation trajectory

2. Arc interpolation of point by point comparison method

1) Principle of point by point comparison method

The step of circular-arc interpolation is the same as that of linear interpolation.

Taking the NR_1 of the circle center as an example, the principle of point by point comparison arc interpolation is explained:

(1) Deviation judgement

The starting point is (x_0, y_0), the end point is (x_e, y_e), the interpolation point is (x_i, y_i), as shown in Figure 2-15, the circle center is at the origin, the radius of the general expression of R is $x^2 + y^2 = R^2$.

The deviation function is $f_i = x_i^2 + y_i^2 - r^2$.

If $f_i = 0$, then the interpolation point (x_i, y_i) is just in the arc;

If $f_i > 0$, then the interpolation point (x_i, y_i) is outside the circle;

If $f_i < 0$, then the interpolation point (x_i, y_i) is inside the circle.

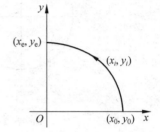

Figure 2-15 Schematic diagram of the counterclockwise arc

(2) Feed control

When $f_i > 0$, then the step will feed toward the $-x$ direction;

When $f_i \leqslant 0$, then the step will feed toward the $+y$ direction.

(3) New deviation calculation

The recursive formula for the error function is as follows.

If a step is in the direction of $-x$, then

$$f_{i+1} = (x_i - 1)^2 + y_i^2 - r^2$$
$$= x_i^2 - 2x_i + 1 + y_i^2 - r^2$$
$$= f_i - 2x_i + 1$$

Similarly, if toward y direction, then

$$f_{i+1} = x_i^2 + (y_i + 1)^2 - r^2$$
$$= x_i^2 + y_i^2 + 2y_i + 1 - r^2$$
$$= f_i + 2y_i + 1$$

The recursion formula shows that the current coordinates of interpolation points should be recorded in the interpolation process in real time.

(4) End-point discrimination

The interpolation stops when the current coordinate is the same with the end-point coordinate. Other method maybe causes error (误判). For example, for the whole circle, the starting point and the end point coincide, if the judge method similar with straight line is used, the interpolation calculation of the circle will not be carried out.

When the circle center is not at the origin, it can be translated to the origin. The method is same.

2) Circular interpolation feed direction

Four quadrant arc feed directions are shown in the Figure 2-16.

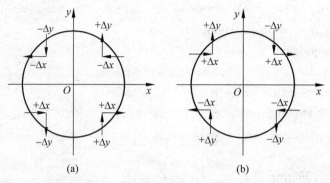

Figure 2-16　Four quadrant arc feed directions
(a) Counterclockwise arc; (b) Clockwise arc

3) Case verification

Example: Set the starting point of the arc at (8, 0), the end point at (0, 8), the center at (0, 0).

For counterclockwise(逆时针的) circular interpolation, the resulting interpolation points and deviations are shown in Table 2-4.

When the deviation is zero, it will feed toward $+y$. The corresponding interpolation pattern is shown in Figure 2-17. Figure 2-18 is for the minimum deviation of the interpolation trajectory. The two figures show that the minimum deviation of the two axes(轴) can be fed at the same time, so the interpolation accuracy is higher and faster.

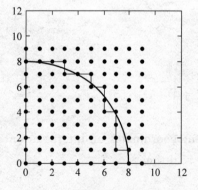

Figure 2-17　Starting point (8, 0), end point (0, 8), center (0, 0) arc point comparison method interpolation trajectory

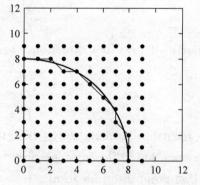

Figure 2-18　Starting point (8, 0), end point (0, 8), center (0, 0) arc minimum deviation method interpolation trajectory

Table 2-4 Arc interpolation calculating

Serial number	Deviation judgement	Feed control	New deviation calculation	End point discrimination	Current end
0			$f_0=0$	(0, 8)	(8, 0)
1	$f_0=0$	$+\Delta y$	$f_1=f_0+2y_0+1=1$		(8, 1)
2	$f_1>0$	$-\Delta x$	$f_2=f_1-2x_1+1=-14$		(7, 1)
3	$f_2<0$	$+\Delta y$	$f_3=f_2+2y_2+1=-11$		(7, 2)
4	$f_3<0$	$+\Delta y$	$f_4=f_3+2y_3+1=-6$		(7, 3)
5	$f_4<0$	$+\Delta y$	$f_5=f_4+2y_4+1=1$		(7, 4)
6	$f_5>0$	$-\Delta x$	$f_6=f_5-2x_5+1=-12$		(6, 4)
7	$f_6<0$	$+\Delta y$	$f_7=f_6+2y_6+1=-3$		(6, 5)
8	$f_7<0$	$+\Delta y$	$f_8=f_7+2y_7+1=8$		(6, 6)
9	$f_8>0$	$-\Delta x$	$f_9=f_8-2x_8+1=-3$		(5, 6)
10	$f_9<0$	$+\Delta y$	$f_{10}=f_9+2y_9+1=10$		(5, 7)
11	$f_{10}>0$	$-\Delta x$	$f_{11}=f_{10}-2x_{10}+1=1$		(4, 7)
12	$f_{11}>0$	$-\Delta x$	$f_{12}=f_{11}-2x_{11}+1=-6$		(3, 7)
13	$f_{12}<0$	$+\Delta y$	$f_{13}=f_{12}+2y_{12}+1=9$		(3, 8)
14	$f_{13}>0$	$-\Delta x$	$f_{14}=f_{13}-2x_{13}+1=4$		(2, 8)
15	$f_{14}>0$	$-\Delta x$	$f_{15}=f_{14}-2x_{14}+1=1$		(1, 8)
16	$f_{15}>0$	$-\Delta x$	$f_{16}=f_{15}-2x_{15}+1=0$	(0, 8)	(0, 8)

3. The accuracy of point by point comparison interpolation

The accuracy is not more than one pulse equivalent, and the proof process is found in "interpolation principle" edited by Li Enlin[10].

2.2.3 Digital differential analyzer interpolation

Digital differential analyzer (DDA) interpolation not only can be easily achieved in the straight line or the curve of second order, also for a variety of computing functions, and easy to realize coordinate linkage(协调联动), so DDA interpolation is used widely.

<small>xié tiáo lián dòng</small>

1. Linear interpolation of numerical integration method

1) Linear interpolation principle of numerical integration method

There is a straight line in the plane OA, the starting point is the coordinate origin O of the end point is $A(x_e, y_e)$, the equation of the straight line is as follows:

$$y = \frac{y_e}{x_e}x$$

Its parametric equation with t is:

$$x = kx_e t, \quad y = ky_e t$$

Where k is the scale factor.

Then the differential equation is obtained:
$$dx = kx_e dt, \quad dy = ky_e dt$$
And then integrating:
$$x = \int dx = k\int x_e dt, \quad y = \int dy = k\int y_e dt$$
If the integrals are expressed in the forms of accumulation, they are approximate to
$$x = \sum_{i=1}^{n} kx_e \Delta t, \quad y = \sum_{i=1}^{n} ky_e \Delta t$$
When $\Delta t = 1$, approximate differential form is:
$$\Delta x = kx_e \Delta t, \quad \Delta y = ky_e \Delta t$$

Δt is interpolation period, kx_e and ky_e are increments at the same time for the two accumulators. The overflow occurs when the accumulated value exceeds a coordinate unit (pulse equivalent). The overflow pulse drives the servo system to feed a pulse until a given line is machined well.

After accumulating m times, x and y reach the end (x_e, y_e), that means
$$x = \sum_{i=1}^{m} kx_e = kx_e m = x_e$$
$$y = \sum_{i=1}^{m} ky_e = ky_e m = y_e$$

Thus, there is a relationship between the proportion of k and the cumulative number:
$$km = 1$$
$$m = 1/k$$
where the value of k and the capacity of accumulator should be greater than the maximum coordinate value. Assume that the accumulator has n-bit wordlength:
$$k = \frac{1}{2^n}$$

Cumulative number: $m = 1/k = 2^n$.

The above results show that if the bit accumulator is up to number n, the whole process of interpolation can be 2^n times to reach the end of the line.

In the register, all the numbers are binary, so kx_e (or ky_e) and x_e (or y_e) is the same, only the decimal point is different. So x_e can be used directly in the x-axis accumulator, using y_e directly in the accumulator y-axis.

When the numerical integration method is used in linear interpolation, each pulse of the accumulator overflowed will drive the corresponding axis. When the accumulated $m = 2^n$ times, the x-axis and the y-axis of the number of steps is exactly equal to the coordinates of the end of each axis.

The end-point discriminant for the linear interpolation is completed by the end-point counter register, whose initial value is 0. The counter will plus 1 with each cumulative(累积的) time, when m times is cumulative, it will overflow. The counter is cleared to 0 again, and the linear interpolation will end. In order to ensure that only one pulse is overflowed at each time, the bits of the counter register should be the same with the register bits of x_e and y_e.

2) Case of numerical integration method for linear interpolation

The linear is OA. The starting point is at the origin O, end point is $A(7, 10)$, the accumulator and register capacity are four bits, the cumulative number is $n = 2^4$. Before the interpolation, the accumulator and registers are 0, registers $x_e = 111$, $y_e = 1010$, respectively, the maximum capacity is $2^4 = 16$. For this interpolation, a numerical integral method is used to calculate and draw the trajectory. Interpolation process is shown in Table 2-5, walking trajectory as shown in Figure 2-19.

Table 2-5 Linear interpolation operation of DDA

Cumulative number n	x integrator		y integrator	
	$J_{Rx}+J_{vx}$	Overflow Δx	$J_{Ry}+J_{vy}$	Overflow Δy
1	0000+0111=0111	0	0000+1010=1010	0
2	0111+0111=1110	0	1010+1010=0100	1
3	1110+0111=0101	1	0100+1010=1110	0
4	0101+0111=1100	0	1110+1010=1000	1
5	1100+0111=0011	1	1000+1010=0010	1
6	0011+0111=1010	0	0010+1010=1100	0
7	1010+0111=0001	1	1100+1010=0110	1
8	0001+0111=1000	0	0110+1010=0000	1
⋮	⋮	⋮	⋮	⋮
16	0000	1	0000	1

If the machined line is shorter, and the wordlength of registers and accumulators is longer, there will be multiple times to generate an overflow (溢出的) pulse and the feed rate will be very slow so as to affect productivity. So, it is possible that the x_e, y_e will also be magnified 2^m times before accumulated, that is to change the location of the overflow pulse to improve the feed rate. But at this point, the end-point judgement should be changed accordingly. For the numbers of x_e, y_e are amplified by 2^m times, the cumulative number should be reduced to $n/2^m = 2^N/2^m = 2^{N-m}$.

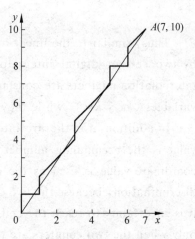

Figure 2-19 Linear interpolation trajectory of DDA

2. Circular arc numerical integration method

Taking the first quadrant inverse circle as an example, as shown in Figure 2-20. The starting point is $A(x_0, y_0)$, the end point is $B(x_e, y_e)$, the circle center is at the the coordinate origin and the radius is R. $P(x_i, y_i)$ is a moving point with the relationship $x_i^2 + y_i^2 = R^2$. The differential equation is

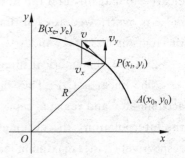

$$2x_i \frac{dx_i}{dt} + 2y_i \frac{dy_i}{dt} = 0$$

$$\frac{dy_i/dt}{dx_i/dt} = -\frac{x_i}{y_i}$$

Figure 2-20 DDA first-quadrant counterclockwise arc interpolation

where $dx/dt = v_x$ is the velocity of the moving point P along the x direction, and $dy/dt = v_y$ is the velocity of the y direction.

The parameter equation likes this:

$$\frac{dx_i}{dt} = -ky_i, \quad \frac{dy_i}{dt} = kx_i$$

where k is the scale factor.

The definite integral can be gotten from the point A to point B. The t_0 and t_n correspond to the time of the starting and the end points, respectively:

$$x_e - x_0 = -\int_{t_0}^{t_n} ky_i dt, \quad y_e - y_0 = \int_{t_0}^{t_n} kx_i dt$$

Replace the integral formula with the accumulation, then obtain

$$x_e - x_0 = -\sum_{i=1}^{n} ky_i \Delta t, \quad y_e - y_0 = \sum_{i=1}^{n} kx_i \Delta t$$

If Δt is taken as one pulse interval, $\Delta t = 1$, then

$$x_e - x_0 = -\sum_{i=1}^{n} ky_i, \quad y_e - y_0 = \sum_{i=1}^{n} kx_i$$

Thus, similar to the linear interpolation, circular interpolation can also be achieved by two sets of digital integrators. The difference between them is that the linear interpolation numbers are constants (kx_e, ky_e), but the arc interpolation numbers are variables (kx_i, ky_i), while they will change with the overflow pulse.

In addition, for the arc interpolation, the current value (x_i) of the x-coordinate value is the accumulate number of the y-axis, and the current value (y_i) of the y-coordinate value is used as the accumulate number of the x-axis. At the end of the discrimination, because the two coordinates of the arc interpolation are done separately, it is uncertainty to reach the end point at the same time. For the end-point judgement, only when the two counters are reduced to 0, the DDA interpolation stops.

For different quadrants of the arc interpolation, the calculation process is the same if the absolute coordinate values are used.

The feed direction is as shown in Table 2-6.

Table 2-6 DDA arc interpolation feed direction and the modification of the register

Arc direction	Clockwise				Counterclockwise			
quadrant	I	II	III	IV	I	II	III	IV
$Jvx(y_i)$	−	+	−	+	+	−	+	−
$Jvy(x_i)$	+	−	+	−	−	+	−	+
x-coordinate feed direction	+	+	−	−	−	−	+	+
y-coordinate feed direction	−	+	+	−	+	−	−	+

2.2.4 Sampled data interpolation

Sampled data interpolation is widely used in the modern CNC system. For linear interpolation, the straight line of the moving point in a period coincides with the given line. For the circular arc interpolation, the moving point in the interpolation period approaches the arc by the string. For example, the interpolation cycle is 8 ms, the position feedback sampling period is 4 ms. The sampling period is two times to the interpolation cycle, and bowstring feed line will take the place of the <u>circular arc interpolation curve</u>（圆 弧 插 补 曲 线）, which is called direct function interpolation algorithm. The straight line and circular arc interpolation algorithms will be introduced below.

1. Linear interpolation algorithm

In the Figure 2-21, for the straight line, the starting point is at the origin, while the end point is $P_e(x_e, y_e)$. The angle between the tool moving direction and the long axis is α, and OA is the first interpolation feed step $\Delta L = FT$ (where F is feed rate and T is interpolation period), while the interpolation length of the long axis is obtained:

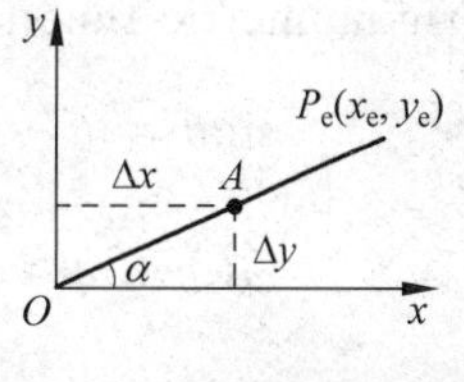

Figure 2-21 Time-division method linear interpolation

$$\Delta x = \Delta L \frac{x_e}{\sqrt{x_e^2 + y_e^2}}$$

The feed of the short axis is derived:

$$\Delta y = \frac{y_e}{x_e} \Delta x$$

Then

$$x_{i+1} = x_i + \Delta x$$
$$y_{i+1} = y_i + \Delta y$$

The interpolation will not end until the moving point is coincided with the end point or the distance between the two points is within the tolerance.

2. Recursion function method for circular interpolation

In the case of the first quadrant, for example, as shown in Figure 2-22, the starting point of the circle is $A_0(x_0, y_0)$, the end point is $A_e(x_e, y_e)$, the radius is R, arc center is located at the coordinate origin, and feed rate is F. $A_i(x_i, y_i)$ is the current point. After one period of T, $A_{i+1}(x_{i+1}, y_{i+1})$ will be reached. The tool motion path is P_iP_{i+1}, while θ is the passing angle called step angle, $\theta = \dfrac{FT}{R} = k$. ϕ_{i+1} and ϕ_i are separately the angle between A_{i+1}, A_i and x-axis. So the following formula can be gotten as

$$\begin{cases} x_i = R\cos\phi_i \\ y_i = R\sin\phi_i \end{cases}$$

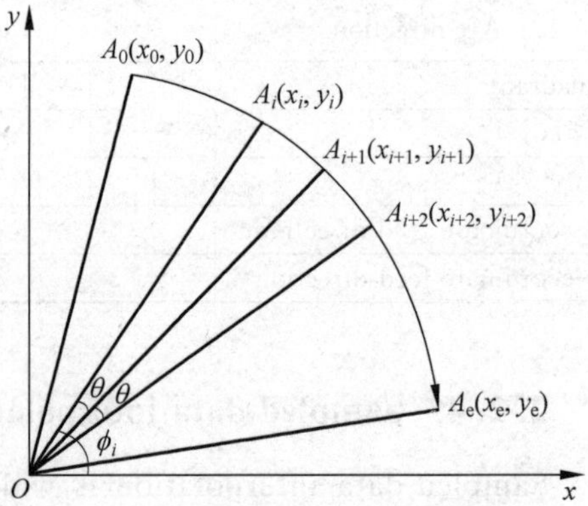

Figure 2-22 The sampled-data interpolation for arc in the first quadrant

After one step interpolation, $\phi_{i+1} = \phi_i - \theta$, then

$$\begin{cases} x_{i+1} = R\cos(\phi_i - \theta) = x_i\cos\theta + y_i\sin\theta \\ y_{i+1} = R\sin(\phi_i - \theta) = y_i\cos\theta - x_i\sin\theta \end{cases}$$

Above formula is called first-order recursion interpolation formula.

Trigonometric functions $\sin\theta$ and $\cos\theta$ will be expressed by the power series:

$$\sin\theta = \theta - \frac{1}{3!}\theta^3 + \cdots + (-1)^n \frac{x^{2n+1}}{(2n+1)!} + \cdots \approx \theta = k$$

$$\cos\theta = 1 - \frac{\theta^2}{2!} + \cdots + (-1)^n \frac{x^{2n}}{(2n)!} + \cdots \approx 1 - \frac{\theta^2}{2} = 1 - \frac{k^2}{2}$$

Then the simplified recursion formula can be gotten as follows

$$\begin{cases} X_{i+1} = x_i + k\left(y_i - \dfrac{1}{2}kx_i\right) \\ Y_{i+1} = y_i - k\left(x_i + \dfrac{1}{2}ky_i\right) \end{cases}$$

The interpolation will not end until the moving point coincides with the end point or the distance in-between is within the tolerance.

2.2.5 Acceleration and deceleration control

If the interpolation control algorithm ignores(忽略) the speed control, then the interpolation can be used only in computer graphics. And the acceleration/deceleration control and interpolation algorithm need to be combined to form a complete CNC system motion control module. In the pulse-incremental interpolation algorithm, the feed rate can be controlled by changing the interpolation period T. In the sampled-data algorithm, the feed rate(进给速度) is not directly related to the interpolation period. The acceleration/deceleration control is divided into the one before interpolation and the one after interpolation. Since the post-acceleration/deceleration mode is considered separately for each axis, in which not only the machining accuracy is lost but also the end-point discrimination error may be caused. The former acceleration/deceleration mode is adopted in the high-precision machining, however, it is very difficult to predict the deceleration point.

Acceleration/deceleration control methods are divided into trapezoidal (Figure 2-23), exponential (Figure 2-24), parabolic and composite curve types. The linear acceleration/deceleration method is simple, but there is adverse impact. The exponential method has no impact, but the speed is slower than the straight line one, and the calculation is complicated. The composite curve acceleration/deceleration method does not have the impact and the speed is moderate, but the calculation is also complicated. It is important to select the appropriate acceleration/deceleration control method according to the different control precision requirements.

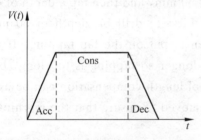
Figure 2-23　Trapezoidal acceleration and deceleration

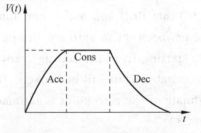
Figure 2-24　Index acceleration and deceleration

2.3　Tool compensation principle

Tool compensation is known as tool offset. In 60—70 years of 20th century, there was no compensation, so the programmer had to focus on the relationship between the theoretical path(理论路径) of the workpiece and the actual tool center. Once the tool radius changes, the tool center must be calculated manually. The efficiency is very low. So, the concept of compensation has greatly improved the efficiency of programming.

During the preparation for the program, if there is tool compensation function in the CNC system, only the theoretical contour programming parts should be considered. But the actual tool radius and length as tool compensation parameters are put into the CNC system, the required workpieces will be gotten.

Tool compensation can also meet the processing technology and other requirements.

Both the rough and the fine machining can be realized by gradually changing the tool radius compensation value. Furthermore, due to tool wear, re-grinding and the tool length changed, the original program is bound to cause processing errors, but the tool compensation(刀具补偿) can solve this problem.

Tool compensation includes tool length compensation and tool radius compensation.

2.3.1 Tool length compensation

1. Conception of tool length compensation

Tool length compensation is very important which can be used for the compensation of the tool axial (Z) direction. If tool length changes, it is unnecessary to change the program, i.e. only to regulate the amount of offset. It should be noted that the control point for the milling machine or the machining center in the Z direction is a fixed point within the tool holder, which is pre-set by the machine tool manufacturer(制造商), but the tool tip is used to machine workpieces, so there is an offset between the control point and the machining point, which can be amended by the length compensation.

For example, to drill a hole with a depth of 50 mm, and then tap a depth of 45 mm, with a 250 mm drill and a 350 mm long tap. First to drill bit depth of 50 mm, the machine has been set to zero for the parts, then to put on the tap tapping. If the two tools are starting from the setting zero, tap is longer so tapping is too long, both the workpiece and the tap will be broken. If the tool length compensation can be used, the Z-coordinate of the zero point is automatically moved to ensure that the machining zero is correct.

Using the tool length compensation command, it is unnecessary to take into account the actual length of the tool. When the tool length is changed due to tool wear, tool change, etc., it is necessary to correct only the tool length compensation amount without adjusting the program or tool.

2. Tool length compensation command

G43 is positive compensation, that is, Z-coordinate word pluses the length of the amount of compensation saved in H code and Z value in the G54, then to drive Z-axis to move by the sum.

G44 is negative compensation, that is, Z-coordinate size word minus H length.

G49 is to cancel the tool length compensation.

Each tool has its own length compensation, when the tool is changed, the use of G43 (G44) H command will give its own tool length compensation and automatically cancel the length of the previous tool compensation.

3. Tool length compensation method

As shown in Figure 2-25, the upper surface of the blank is the Z_0 in the workpiece coordinate system. Since the control point in the Z direction is a fixed-point B in the tool holder, not the tool nose point, when the tool length is different, the mechanical coordinators of B point are different, that is, Z value in the G54 is also different. Here are two types of tool length compensation.

(1) The actual length of the tool is used as the compensation value for the tool length. The method is to use the tool length detector to measure the length of the tool, and then enter this value into the tool length compensation register as the compensation value, and then select one tool as a reference tool to measure the Z value of the workpiece surface, which can be input into the Z reference value of G54.

This method is applicable for many large machinery processing enterprises. For those companies specialized in tool management, every tool has the parameters file. The operator can use the tool through the tool file, and unnecessarily to measure the tool length. Operation and measuring can be run separately, so improve the efficiency.

(2) Use the Z-direction distance between the tool tip and the programmed zero as the compensation value. This method is suitable for only one person to operate simple workpiece and not having enough time to use the detector. When this method is used, the Z value in the coordinate offset (G54—G59) is set to 0.000, and the compensation value of the tool length is the offset value of workpiece surface in mechanical coordinate system for different tool, so the compensation value is always negative and large. It is unnecessary to measure every tool length but it is a must to check the Zero coordinate of Z-axis for every tool before machining.

4. Example of tool length compensation

As shown in Figure 2-25, execute the instructions as follows

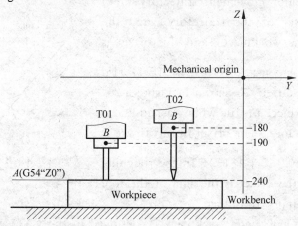

Figure 2-25　Tool length compensation

G90G54G00X0Y0;

G43H01Z0.000.

(1) With the actual length of the tool as the tool length compensation value, then the Z parameter in G54 should be −240.000. In the parameter table H01, enter the tool length of the T01 as 50.000, and enter the tool length of 60 in H02. When G43H01Z0.000 is executed, the tool T01 tip will get the workpiece surface according to 0 + (−240) + 50 = (−190), while the workpiece coordinator of Z is 0.000 and mechanical coordinator is −190. The process is similar to T02.

(2) Use the Z direction distance between the fixed-point B and the programmed zero(编程原点)as the compensation value, then the Z parameter in G54 should be 0.000, in the parameter table H01, −190.000 should be input (for the protection of the workpiece surface, the standard detection block with 100 mm is used to test the compensation. The block is perpendicular to the workpiece surface, and then make the tool tip touch block surface, then get the mechanical coordinate of −90 for Z direction. So −190.000 mm is the compensation value of T01 for workpiece surface in mechanical coordinate system). In H02, enter −180.000. When G43H01Z0.000 is executed, the tool T01 tip will get the workpiece surface according to 0 + (−190) + 0 = (−190), while the workpiece coordinator of Z is 0.000 and mechanical coordinator is −190. Similarly, if the G43H02Z0.000 will be implemented, tool T02 tip also happens to move to the workpiece surface(表面).

For the students to do experiments or training, the second method is more simple and useful commonly.

2.3.2 Tool radius compensation

1. Principle of the tool compensation

In the contour processing, there is an offset between the tool center movement trajectory (tool center or wire center of the trajectory) and the actual contour of the workpiece, and this offset is called tool radius compensation(半径补偿), also known as tool center offset. As shown in Figure 2-26, solid line is the workpiece profile while dashed line is the tool center trajectory. The offset in this figure is the tool radius value(刀具半径值). In the roughing and semi-finishing, the offset is the sum of the tool radius and the machining allowance.

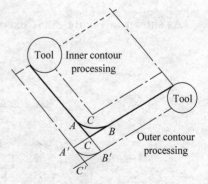

Figure 2-26 B function tool offset and break point

Because only the tool center track can be controlled in the CNC system, this system must calculate the tool center track according to the workpiece contour size and tool radius compensation value. At the tool compensation command, CNC system can automatically compensate tool radius. For manual programming, tool radius compensation is particularly important.

When the tool has a small amount of wear, processing contour size and design size slightly deviate(偏离)or the processing allowance changes during the rough/semi-fine/fine/milling, without modifying the program, only the tool radius compensation value makes appropriate change. Tool radius compensation not only simplifies the programming calculation, but also increases the readability of the program.

There are B function (Basic) and C function (Complete) tool radius compensation. For the B function tool radius compensation, it cannot solve the transition between two sections, shown in Figure 2-26, because tool center path is calculated only by the procedures of this section self. The intermittent points and intersections must be calculated manually. The C tool radius compensation can automatically handle the intermittent points and intersections of the two near blocks. Almost all modern CNC systems have the C function tool radius compensation.

2. Command of tool radius compensation

According to ISO regulations, when the tool center track is located at the left side of workpiece by the forward direction of the tool moving, it is called the left tool compensation and G41 is the command; otherwise called the right tool compensation and G42 is the command.

G41 is the tool left compensation command, that is, along the tool forward direction (assuming the workpiece does not move), the tool center track is located on the left side of the workpiece contour, shown in Figure 2-27(a).

G42 is the tool right compensation command (right knife fill), that is, along the tool forward direction (assuming the workpiece does not move), the tool center track (轨迹) is located on the right side of the workpiece contour, shown in Figure 2-27 (b).

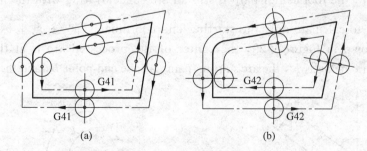

Figure 2-27　Tool radius compensation
(a) Left side tool compensation; (b) Right side tool compensation

G40 is to cancel the tool radius compensation command. When G40 is used, the G41 and G42 instructions are invalid.

When the G41 or G42 is used, the following steps should be taken:

(1) Set the tool radius compensation value. Input this value in the tool compensation parameter area before the program starts.

(2) Establish the tool compensation. The tool is close to the workpiece from the starting point, while the end point of the tool center track is the normal direction offset point of the next block. In this block, G00 or G01 can be used together with G41 or G42.

(3) Run the tool compensation. During compensating, the tool center track always deviates from the workpiece by a tool compensation value. In this state, G00, G01, G02, G03 can be used.

(4) Cancel the tool compensation. Tool compensation is cancelled to return to the origin of the process. In this block, only G00 or G01 can be used together with G40.

3. B function tool radius compensation

For a straight line, the tool center path is a straight line parallel to the original line, so only need to calculate the start-/end-coordinates of tool center track.

As shown in Figure 2-28, the starting point of the processed line segment is at the coordinate origin and the end point is A. Assume that the tool center is known at the point O'. For the tool radius r, it is now necessary to calculate the end-point A':

$$X' = X + \Delta X$$
$$Y' = Y + \Delta Y$$
$$\angle xOA = \angle A'AK = \alpha$$
$$\Delta X = r\sin\alpha = r\frac{Y}{\sqrt{X^2 + Y^2}}$$
$$\Delta Y = -r\cos\alpha = -r\frac{X}{\sqrt{X^2 + Y^2}}$$
$$X' = X + \frac{rY}{\sqrt{X^2 + Y^2}}$$
$$Y' = Y - \frac{rX}{\sqrt{X^2 + Y^2}}$$

For an arc, the tool center path is still an arc concentrating with the machined arc. It is necessary to calculate only the starting/end coordinates(起终点坐标 qǐ zhōng diǎn zuò biāo) of the tool center. As shown in Figure 2-29, the center of the processed arc is at the coordinate origin O, the arc radius is R, the arc starting-point A, the end-point B, and the tool radius r.

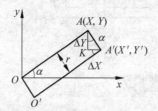

Figure 2-28 Line tool compensation for B function

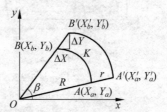

Figure 2-29 Arc tool compensation for B function

Assuming that the tool center is A', its coordinates are known. Then to calculate the coordinates of end B' of the tool center track.

$$X'_b = X_b + \Delta X$$
$$Y'_b = Y_b + \Delta Y$$
$$\angle BOx = \angle B'BK = \beta$$
$$\Delta X = r\cos\beta = r\frac{X_b}{R}$$
$$\Delta Y = r\sin\beta = r\frac{Y_b}{R}$$
$$X'_b = X_b + \frac{rX_b}{R}$$
$$Y'_b = Y_b + \frac{rY_b}{R}$$

In the case of the outer contour of Figure 2-26, the tool center falls at the point B' which can be calculated by the B-function tool radius compensation method, but the starting point of the second block is A'. There is a breakpoint(断点) between the two blocks so the tool center path needs an additional(附加的) path from B' to A' which is finished manually in B-function tool radius compensation. In order to solve this problem, the next block must be read in advance and then the transition trajectory will be calculated and corrected automatically(自动地), which is the C-function tool compensation. C-function tool is more perfect which can handle the two-block tool center track conversion based on the adjacent contour segment, and automatically inserted transition arc or straight line to improve the efficiency and stability of playing program greatly.

4. C-function tool radius compensation

The straight lines and arcs are used widely, so the corresponding transfer includes straight to straight, straight to arc, arc to arc and arc to line.

Before discussing the transition method of the tool radius compensation, the meaning of the vector angle is explained. It refers to the nonprocessing side angle between the two block tracks.

According to the difference between the vector angle of the two trajectories and the direction of the tool compensation, there are three transition modes, including shortened, extended(伸长的) and insert types.

When $\alpha \geqslant 180°$, it is shortened type; when $180° \geqslant \alpha \geqslant 90°$, it is extended type; when $\alpha < 90°$, it is insert type; as shown in Figure 2-30.

In the vector calculation, the method used is plane analytic geometry, rather than solving the simultaneous equations(联立方程). Because its solution is rather complex, in addition, when there are multiple results, it is necessary to single out the only one. The specific solution can be found in ordinary textbooks of advanced mathematics.

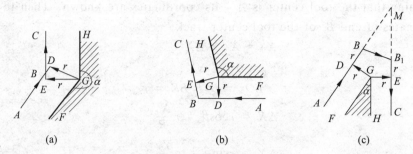

Figure 2-30 Three tool compensation modes
(a) Shortened type; (b) Extended type; (c) Insert type

5. Case of C-function tool radius compensation

To machine the outline of $CDEF$, the tool path is followed from the point A to B, D, E, F, G, A. The tool center track is $AabcdefA$, as shown in Figure 2-31. The processing steps are as follows:

(1) G41 G00 XbYb D01, read BD, $\angle ABD > 180°$, this section is transferred to shorten, then calculated the coordinates, and enter the line segment $abAB$, build left side tool compensation;

(2) G01 XdYd, read DE, $\angle CDE = 90°$, this paragraph is transferred to extended type, then calculate the coordinates of b, and enter the straight line ab;

(3) G01 XeYe, read EF, $\angle DEF < 90°$, this section is transferred to insert type, then calculate the coordinates of c, d, and enter bc, cd straight line;

(4) G01 XfYf, read FG, $180° > \angle EFG > 90°$, this section is transferred to extended type, then calculate the coordinates of e;

(5) G01 Xg Yg, read into GA and G40, $\angle FGA > 180°$, transferred to shorten type, then calculate the coordinates of point f;

(6) G40 G01 XaYa, this block is to cancel the cutter compensation, the tool center moves back to point A directly, then finish the machining.

```
G90G54G00XaYaS800M03;(tool center will get to A point)
G43H01Z100.;
Z5.;
G41G00XbYbD01;(the coordinator of B point is (Xb,Yb))
G01Z-5.F50;
XdYd;
Xe Ye;
Xf Yf;
Xg Yg;
G00Z100.;
G40X0Y0;
M30;
```

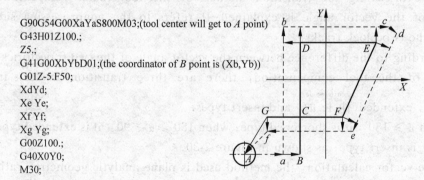

Figure 2-31 Tool radius compensation procedures and graphics

Chapter 3　Servo systems

3.1　Overview of servo systems

3.1.1　Servo system components

In the CNC machine, the servo system(伺服系统) is an execution unit, which drives the worktable, spindle and other moving parts to achieve expected movement. As a comprehensive and complex system, it includes mechanisms, electronic circuits, electric motors (hydraulic motors also used in the early stage), sensors, microprocessors(微处理器) and other components. The servo systems can be classified into feed servo systems and spindle servo systems. The servo system receives the position command from the CNC and utilizes signal processing as well as power amplification to drive the feed motor, realizing the feed motion of the worktable, turret and so on. The spindle servo system drives the spindle motor according to the speed command of the CNC to realize the speed regulation of the spindle. Some spindle servo system using the position closed-loop control(闭环控制) can also achieve spindle positioning, indexing or moving as a normal coordinate feed axis. The accuracy, speed and other parameters of the servo system can directly affect the processing quality and efficiency of CNC machine tools. Therefore, servo system with high performance and high reliability is a key technology of CNC system.

The typical structure of a CNC machine tool feed servo system is shown in Figure 3-1, which includes the control, detection and feedback, power amplification, motor and feed mechanism. The typical servo system consists of three control loops, i.e. the current, the speed and the position loops. The current loop consists of a current controller, a current detection device and a current feedback. The speed loop consists of a speed controller, a speed detection device and a speed feedback. The position loop consists of a position controller, a position detection device and a position feedback. In the movement of the machine tool, the controller generates control signals according to the position command and the feedback of the detection device, which involves the signal conversion and power amplification to drive the motor and feed mechanism to achieve movement and position command tracking. The spindle servo system is similar to the feed one, except that there is usually no position control, detection and feedback(反馈) and that it can only control the speed of the spindle. When the spindle needs to achieve quasistop(准停), indexing or rigid tapping functions, the spindle position control must be included. In addition, it should be noted that, in the CNC machine tools, the good synchronization(同步) among coordinated axes is needed to ensure the accuracy of contour machining, as well as the good dynamic and static performance.

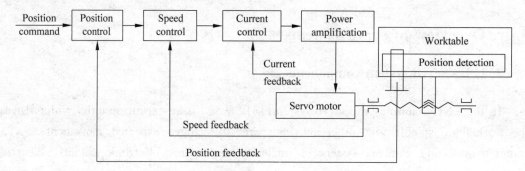

Figure 3-1 General structure of CNC machine tool servo system (fully closed-loop)

3.1.2 Servo system classification

1. Classification by the control principle

1) Open-loop servo system(开环伺服系统)

As shown in Figure 3-2, the structure of the open-loop servo system does not contain position detection and feedback parts. In CNC machine tools, such system is generally driven by the stepper motor which converts a command pulse from the NC control to a rotation angle. Command pulse is converted by the frequency conversion, pulse distribution, power amplifier circuits to drive the stepper motor to rotate, which drives the gears, ball screw nut pair and guide rails to achieve the motion of the worktable. The number of pulses determines the distance of the worktable movements, and the frequency of the pulses determines the speed of the movement. The open-loop servo system is simple in structure, low in cost and easy for tuning and maintenance.

However, the angular displacement error of the stepper motor, the errors of the gears, screw and other transmission parts(传动部件), as well as the mechanism gaps and frictions will affect the movement accuracy, and these errors cannot be detected by the feedback device for compensation(补偿), so the accuracy of open-loop servo system is fairly low. Besides, the low-speed motion of the stepper motor is not smooth enough while the high-speed torque output is small, thus resulting in losing steps. Hence, the open-loop servo system is mainly used for low-cost CNC machine tools which do not demand high performance.

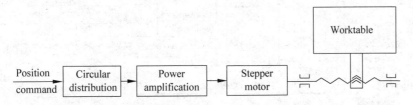

Figure 3-2   Open-loop servo system

2) Fully closed-loop servo system(全闭环伺服系统)

The structure of fully closed-loop control (sometimes also called a closed-loop control system) is shown in Figure 3-1. Compared with the open-loop servo system, it adds the position, speed and current detection devices as well as feedback parts. The position detection device of the fully closed-loop servo system is located on the worktable. The servo system can directly detect the displacement of the table. By comparing to the position command, the fully closed-loop servo system can compensate the position error in the whole kinematic chain, thus improving the positioning accuracy and tracking accuracy. Because of the existence of gaps in the transmission chain(传送链), friction and other nonlinear factors affect the stability of the closed-loop system, the system is prone to oscillation(振荡). The design and tuning of fully closed-loop servo systems is usually complex and costly, and therefore mainly used for precision machining tools.

3) Semiclosed-loop servo system(半闭环伺服系统)

The structure of the semiclosed-loop servo system is shown in Figure 3-3, and its position detection device is located at the intermediate moving part of the feed chain (e.g., on the servo motor shaft). The servo system obtains the position of the worktable indirectly based on the detected(检测) position of the intermediate moving part. Since the feedback device is in the intermediate of the drive chain(驱动链), it is impossible to detect the error between the motor shaft and the worktable (e.g., the error of coupling, screw deformation, pitch error, backlash, etc.). Therefore, the accuracy of the semiclosed-loop servo system is lower than that of the fully closed-loop servo system. However, since the gap, friction and other nonlinear factors are excluded in the control loop, the system is easy to adjust and much more stable(稳定的). Position detection device is generally installed in the servo motor when manufacturing, which simplifies the assembling and adjustment of CNC machine tools. Thus, this structure is widely used in small-and medium-sized CNC machine tools.

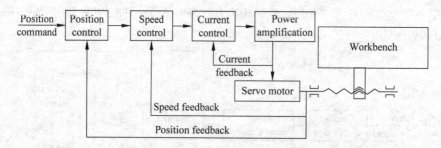

Figure 3-3　Semiclosed-loop servo system

2. Classification by functions

1) Feed Servo System

The feed servo system is used to control the feed motion of the worktable(工作台) or the cutter(刀具) of the CNC machine tools. The contour machining accuracy and quality of the tools depend largely on the performance of servo system.

2) Spindle servo system

The spindle servo system is used to control the rotation(回转) of the spindle of the NC machine tools, which realizes the speed regulation function. For the servo spindle with position control, it also includes the position closed-loop control function.

3. Classification by the feedback comparison control methods

1) Digital pulse comparison servo system

The servo system of CNC will directly compare the generated digital (or pulse) position command with the detected digital (or pulse) signal, and calculate the position error to realize the semiclosed-or fully closed-loop control of the machine tool. Because of its simple structure, stable operation and high reliability(可靠性), this type of servo system is widely used.

2) Phase comparison servo system

In this type of servo system, the position detection device is in the phase operation mode. The position command of the NC system and the position feedback of the detecting device are converted into the phase of a carrier wave. The phase comparison servo system is suitable for inductive detection elements (resolver, inductosyns, etc.). The signal is modulated by high-frequency carrier, which has the characteristics of strong antidisturbance(抗干扰)ability and fast response.

3) Amplitude comparison servo system

In the amplitude comparison servo system, the command position and position feedback are converted into the amplitude(振幅) of the signal. This type of servo system is difficult to realize high-precision control in the large moving range.

4) Full digital control servo system

With the development of digital electronic technology and computer technology, the servo system has begun to develop towards the full digitalization(数字化). In the fully digital servo control system, feedback control of the current, speed and position loops are implemented by digital control using a dedicated microprocessor for real-time processing. This type of control has finished the transition from analog control or hybrid control to full digital control. Full digital control not only improves the accuracy and quality of the servo system, but also provides high-degree flexibility, which can realize a wide range of advanced and intelligent control algorithms, therefore has gradually become the mainstream(主流)of the servo systems for CNC machine tools.

3.1.3 Basic requirements of servo systems

1. Requirements for feed servo systems

1) High precision

The servo system needs to follow the input position command signal accurately. In order to realize the high-precision NC machining, the servo system needs not only the high positioning accuracy, but also the high dynamic tracking accuracy. In general, the positioning accuracy of CNC machine tools should achieve 0.01—0.001 mm, or even 0.1μm. In the multi-axis contour machining, it is necessary to coordinate the movement of each feed axis with the output coordinates of the interpolator to achieve high contour accuracy(精度).

2) Fast response

Response speed is an important index to reflect the dynamic quality of servo system, which affects the tracking accuracy of servo system. In order to achieve a good dynamic response quality, on the one hand, the transient time should be as short as possible, generally within 200 ms, or even down to tens of ms(毫秒); On the other hand, the overshoot should be minimized. These two requirements are often contradicted, which need to be balanced according to the requirements of the process.

3) Wide speed regulation range

Since the demand of the workpiece materials, cutting tools and machining process of CNC machine tools varies in a wide range, the feed speed should be varied in a large range in order to achieve good processing quality and efficiency in different conditions. The speed regulation range is usually defined by the ratio of the minimum feed rate and the maximum feed rate of the servo system, which is generally more than 1 : 10,000. If this value is up to 1 : 24,000, most requirements of the CNC machine tools can be met.

In addition, in the low speed feed motion, the servo system should provide large torque to prevent creeping phenomenon. At zero speed (i.e. table/tool is rest), the motor should provide enough electromagnetic torque to overcome the external disturbance and maintain the positioning precision of the system.

4) Good stability

The stability of the servo system means that the system should achieve a new or restore the original equilibrium state after a short adjustment process under the given input or external disturbance. The servo system should have a strong interference suppression (抑制 yì zhì) capability, and ensure a smooth and steady feed motion in the cutting process. Stability directly affects the machining accuracy and surface roughness (表面粗糙度 biǎo miàn cū cāo dù). In addition, the servo system should also have good robustness, that is, when the parameters of the system (such as workload, ambient temperature, etc.) varies in a certain range, it should still maintain good control quality.

5) High reliability

The servo system of CNC machine tool needs to ensure high-reliability operation in the complex environment (such as voltage(电压 diàn yā)/temperature/humidity fluctuation and vibration, electromagnetic interference, frequent starting/stop). The reliability of the system can be measured by the mean time between failures (MTBF). The larger this value, the higher the reliability.

6) High energy efficiency

In the field of industry, the motor energy consumption accounts for a large proportion of the total. Therefore, it is of great significance to improve the energy efficiency of the servo system to reduce operating costs and carbon emissions and to achieve sustainable development. The method to improve energy efficiency includes using energy-efficient motor materials and electronic components, reducing mechanical friction and adopting efficient processing technology, etc. In addition, some advanced servo system can absorb partly the energy (能量 néng liàng) generated by motor braking, recycling or feedback to the grid.

2. Requirements for the spindle servo system

1) Large output power

The spindle servo system needs to provide enough cutting torque at the specified spindle speed, so it should have a larger output power to meet the requirements of various processes. For CNC machine tool, usually the low spindle speed is used in heavy cutting and rough machining, while the high spindle speed is used for light cutting and finishing machining. Hence the spindle should have a wide range of constant power output.

2) Wide speed range

The spindle speed of the NC machine tool is automatically programmed according to the programming instructions, which should realize stepless speed regulation(调节)in a wide range to reduce intermediate transmission parts and simplify the structure of spindle head.

3) Locating, quasistop and synchronous control

The spindle servo system mainly works in the speed control mode, but in some cases the position control is also needed. For example, in the tool exchange process of the machining center, the quasistop function is needed to facilitate going in/out of tool magazine in the correct direction. The indexing function of turning centers also requires the C-axis to be stopped at a given angle. In rigid tapping, the spindle and the feed axis should be synchronized to the position command of the interpolator for thread machining. The position control function requires the spindle servo system to add the encoder for position detection and realize the position closed-loop.

In addition, the spindle servo system also requires the control accuracy and response speed(响应速度), good stability, reliability and high efficiency.

3.2 Commonly used driving elements

3.2.1 Stepper motors

Stepper motor, also spelled as "stepping motor" or "step motor", is a kind of electromagnetic incremental motion actuator(电磁增量运动执行器), which is mainly used in the open-loop servo system of CNC machine tools. The position pulse signal sent by the CNC is processed through the circular distribution and the power amplifier to drive the stepper motor, which generates an angular displacement to drive the screw nut pair(丝杠螺母副) and other transmission mechanism so as to convert the angular displacement to the linear movement of the worktable. Each pulse corresponds to an angular displacement of the stepper motor, so the stepper motor works in a full NC mode, which is very convenient to interface with a microprocessor/microcomputer. The stepper motor(步进电机) is generally applicable to small-and medium-sized CNC machine tools. At present, there are some stepper-motor-driven CNC machine tools which uses position detection components to construct a closed-loop control system with position feedback, and can achieve a higher control performance.

1. Classification of stepper motors

According to the principle of operation, stepper motors can be divided into 3 types: reactive, excitation and hybrid steppers. The rotor of the reactive stepper motor has no windings, which are made of soft magnetic materials (i. e. easy magnetization/

demagnetization). The reactive stepper motor has the advantages of simple structure and small step angle, but the efficiency is low, the heat is large, and the dynamic performance is poor. The rotor of the excitation stepper motor is made of permanent magnetic material (or excited by windings), therefore the control power is small, the efficiency is high, the damping is large and the dynamic performance is much better. Furthermore, the excitation stepper can hold a certain torque even when power off. However, the step angle of the permanent magnet stepper motor is large (more than 5°), therefore it is rarely used in CNC machine tools. The rotor of the hybrid stepper motor, also known as the permanent-magnet induction-type rotor (永磁感应式转子), is composed of a permanent magnet in the middle and a soft magnetic material at both ends. At present, the hybrid stepper motor is gradually becoming the mainstream of the stepper motors in CNC machine tools.

In addition, according to the output torque of the stepper motors, they can be divided into servo type (can drive only small loads) and power type (drive only CNC machine tools and other heavy load). According to the number of phases, they can be divided into the three-phase, the four-phase, the five-phase, the six-phase, etc. According to the configuration of windings, they can be divided into radial phase division and axial phase division. According to the motion types, they can be divided into rotary (旋转的), linear, plane and rolling motion types. According to the number of stators, they can be divided into single-, double-, triple-or multiple-stator types.

2. The structure and working principle of stepper motors

Because the permanent magnet stepper motors are rarely used in CNC machine tools, and hybrid (混合的) stepper motors and reactive stepper motors are very similar, so we take the reaction stepper motors as an example to illustrate the structure and working principle of step motors. The structure of a typical single-stator, radial-phase-divided three-phase stepper motor is shown in Figure 3-4. The stepper motor is composed of a stator and a rotor. The stator part comprises an iron stator core (定子铁芯) and stator windings, where the iron core is a stack of silicon steel sheets, and the windings are coils which are wound on

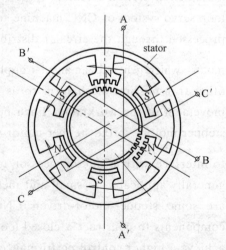

Figure 3-4 Structure of the three-phase reactive stepping motor

the six evenly distributed teeth of the iron core. In the diameter direction, the two opposite teeth are wound by a same coil to form a phase of control winding, which can form a pair of NS poles as shown in Figure 3-4. These poles have 3 pairs, thus the device is called three-phase stepper motor. In the three-phase stepper motor, the directions of the magnetic poles of each phase are differed in 120° in space. On each stator pole surface, there are five equally distributed small teeth facing to the rotor. The corresponding central angle between adjacent teeth is 9°. The rotor is made of soft magnetic material and has no windings. The surface of the rotor has 40 uniformly distributed small teeth. The width of these teeth are the same as on the stator, and the resultant angular distance of the teeth is 360°/40 = 9°(same as the stator). According to the structure, it can be seen that each phase of the stator of the three-phase stepper motor has staggered 1/3 pitch to the rotor in the circumferential direction, i.e. 3°. As shown in Figure 3-5, spreading the stator and rotor(转子) teeth along the circumferential direction into a straight line, assuming the teeth on the stator and rotor of the A-phase are aligned, in the B-phase (120° away from A-phase) the teeth of rotor will be ahead of the teeth of stator by 1/3 pitch (3°), and in the C-phase, the teeth of rotator will be ahead of the teeth of stator by 2/3 pitch (6°). The misaligned condition of the stator and rotor teeth is called staggered teeth, the staggered teeth are a precondition for the operation of the stepping motor. The rotation angle of each moving step is called the step angle, which is related to the angle of the teeth misalignment. The greater number of teeth on the rotor or number of magnetic poles, the smaller the angle of the teeth misalignment and step angle, which results in a higher positioning accuracy and more complex structure of the stepper motor.

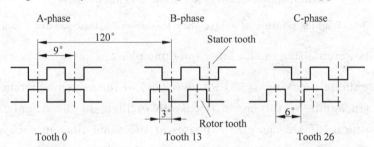

Figure 3-5 The distribution of the relative pitch when A-phase is aligned

If powering the misaligned phases, the rotor will rotate an angle by the electromagnetic force(电磁力) to align the small teeth of rotor and stator. This is the fundamental principle of the movement of stepper motors. The rotation of the stepper motor when power on is illustrated in Figure 3-6. For convenience, we assume that the stator has three phases and each phase has only one tooth, and the rotor has 4 teeth. The first state is that A-phase is energized and B, C phases are de-energized. The first and third teeth of the rotor will move towards the teeth of A-phase of stator through the least reluctance path, and finally stopped when aligned with A-phase. Then switch to the second state to energize the B-phase and de-energize the A, C phases. The magnetic field generated by the B-phase will drive the rotor to rotate 30° in the counterclockwise direction through the minimum reluctance path, and the second and fourth teeth will align with B-phase of the stator. Next, in the third state, energizing the C-phase and de-energizing A and B phases, the rotor continues to rotate counterclockwise 30°, the first and third teeth will be aligned to the teeth of C-phase. According to the above on sequence, the stator is energized with A-B-C-A… phase, and the rotor moves in the counterclockwise direction, in each state switching the rotated step angle is 30°. If the energizing sequence becomes A-C-B-A…, the rotor will rotate in the clockwise(顺时针旋转 shùn shí zhēn xuán zhuǎn)direction, and the step angle is also 30°. In the above mode, there is only one phase that is energized in each state, the number of steps in a cycle is 3, therefore it is called single-powered 3-step mode for three-phase step motors. It can be seen that the number of state switching determines the angular displacement, and the frequency of the state switching determines the angular velocity(角速度 jiǎo sù dù) of the motor.

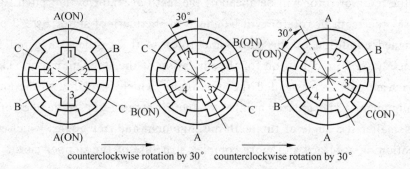

Figure 3-6　Working principle of a three-phase step motor in single-powered 3-step mode

In single-powered 3-step mode, since only one phase is energized in each state, it is possible to lose the self-locking(自锁 zì suǒ) torque(扭矩 niǔ jǔ) at the switching moment and prone to lose steps. In addition, this mode is easy to oscillate in the neighborhood of the equilibrium position. If we energize 2 phases at the same time in each state, it will become a double-powered 3-step mode. In this mode, the counterclockwise energizing sequence is AB-BC-CA-AB…, and the clockwise energizing sequence is AB-CA-BC-AB… In this mode, the magnetic field(磁场 cí chǎng) distribution of each state is different from that of the single-powered 3-step mode, i.e. the teeth of the rotor and the stator are not aligned. For example, when A, B phases are energized, the teeth of the rotor will stagger away ±1/6 teeth pitch from the teeth of poles of A, B phases respectively, and stagger 1/2 teeth pitch from the C-phase. The step angle in the double-powered 3-step mode is also 30°, but the energizing time for each phase is two times of the single-powered 3-step mode, thus it will consume(消耗 xiāo hào) more power and generate more torque. Besides, this mode ensures that at least one phase of the stator winding is energized, so it is not prone to lose steps.

If we use single-powered and double-powered mode in a combined pattern, i.e. energizing with A-AB-B-BC-C-CA-A… sequence, the rotor will rotate in counterclockwise of 15° in each step, which is called 6-step mode for 3-phase motor. Similarly, with the A-CA-C-BC-B-AB-A⋯ energizing sequence, the motor can rotate clockwise. Compared to the 3-step mode, the 6-step mode has half of the step angle, so the step accuracy is doubled. In addition, the 6-step mode also ensures that at least one phase is powered when the state switching, therefore the movement is relatively smooth.

3. Stepper motor characteristics

1) Step-angle and static step error

The step angle α is related to the number of phases of the stator windings, m, the number of the teeth of the rotor, z, and the energizing sequence, which can be written as follows:

$$\alpha = \frac{360°}{mzk} \tag{3.1}$$

where k is the ratio of the number of steps to the number of phases in the energizing sequence. If there are m steps and m phases, $k = 1$; If there are $2m$ steps and m phases, $k = 2$.

The step angle is an important parameter to reflect the accuracy of the stepper motor. Generally speaking, the smaller the step angle, the smaller the pulse equivalent(脉 冲 当 量)and the higher the machining accuracy. The general reaction type and hybrid type stepper motors have step angles between 0.5°—3°. The static step error refers to the difference between the theoretical step angle and the actual step angle when no load is applied (also known as single-step error). The maximum error between the actual angle and the theoretical value of the stepper during a full revolution is defined as the accumulative error, which is the accumulative value(累积值) of the static step error. The step error has a direct influence on the accuracy of NC machine tools.

2) Static torque and the torque-angle characteristics

When energizing the stator, the rotor will arrive at a stable equilibrium position (平衡位置) by the reluctance, and the electromagnetic torque will be zero. If the rotor is moved away from the balance(平衡) point by a torque M applied on the motor shaft, which causes the rotor moving away from of the equilibrium position by an angle of θ, the rotor will be subjected to an electromagnetic torque M_j to balance the load torque M. M_j is known as the static torque and θ is called the misalignment angle. The curve of the relationship of M_j and θ is called the torque-angle characteristic of the stepper motor. If θ is within a certain range, the rotor can be restored to the original stable equilibrium position after removing the external load torque, which is called the static stability zone. As shown in Figure 3-7, it is assumed that a single phase is energized and the pitch angle of the teeth is θ_S, when $\theta = 0$ the motor is in a stable equilibrium point, when $\theta = \pm \dfrac{\theta_S}{2}$ the motor is in an unstable equilibrium point. Figure 3-8 is the torque-angle characteristic

curve(特征曲线), in which the maximum value of the electromagnetic torque, called the maximum static torque $M_{\max}$, can be reached in the stable region.

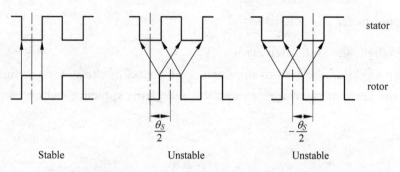

Figure 3-7　Static stable point and unstable point

3) Maximum starting frequency

The maximum starting frequency(频率) refers to the maximum frequency for the stepper motor to start suddenly and not lose steps in a given driving mode. If the starting frequency is greater than the maximum starting frequency, the stepper motor cannot work properly. The maximum starting frequency of stepping motor with load is lower than that of the no-load, and the frequency will decrease with the increment of load.

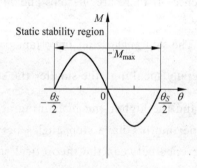

Figure 3-8　Static torque-angle characteristic of the stepper motor

4) Maximum operating frequency for continuous operation

This is the maximum operating frequency for the stepper motor to run continuously without losing steps after starting and accelerating. The frequency determines the maximum speed of the stepper motor.

5) Acceleration/deceleration characteristics(加速减速特性)

If a speed higher than the maximum start speed is required, the control frequency must be gradually increased according to a certain acceleration curve to avoid losing or adding steps. Similarly, in the condition to stop the motor from the working speed, it is necessary to reduce the control frequency in accordance with a certain deceleration curve. The commonly used acceleration/deceleration curves(加速减速曲线) include exponential, linear and S-curve types. No matter what type of acceleration/deceleration curve is applied, the acceleration/deceleration time should not be too short. The acceleration/

deceleration characteristic of the stepper motor is usually described by the acceleration/deceleration time constant(常数).

6) Torque-frequency characteristic and dynamic torque

In the continuous and stable operation condition of the stepper motor, the relationship between the output torque and the operating frequency is called the torque-frequency characteristic, as shown in Figure 3-9. The torque(扭矩) corresponding to each frequency is called dynamic torque. The dynamic torque of stepping motor usually decreases with the increasing frequency. This means that the load ability of the stepping motor is decreased with

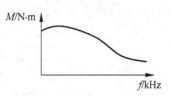

Figure 3-9　Torque-frequency characteristic of stepper motor

the increasing of the working speed, so the torque-frequency characteristic forms an important basis for the selection of step motor.

4. Driving and control for the stepper motors

The stepper motor feed servo system usually works in open-loop control mode, as shown in Figure 3-10. The position command of CNC is sent by the pulse/direction signal to the stepper motor driver. The stepper motor driver includes a circular distributor and a power amplifier. The driver energizes the stator windings(定子绕组) of each phase by a certain sequence to realize position control.

1) Circular distributor

According to the working principle(工作原理) of stepper motors, when the stepper motor runs continuously in one direction, the on/off sequence of each phase is cycled, therefore the process of converting the pulse/direction signal to the on/off signal of each phase is called circular distribution. The circular distribution is not only cycled but also reversible so as to change the direction of motor motion.

The circular distribution function can be realized by either hardware or software. When the hardware circular distribution is implemented(实现), it can be constructed by various kinds of standard flip-flops, or using the special-designed integrated circuit for circular distribution, or using programmable logic devices. The connection between the hardware circular distributor and the NC device is shown in Figure 3-10. The CNC output PULSE and DIR signals to the circular distributor. The PULSE is the step pulse signal. On the rising (or falling) edge of each PULSE signal, the output of circular distributor switches to the next state, the stepper motor rotates a step angle. The DIR signal is the direction signal of the stepper motor, the high or low level corresponding to the clockwise or counterclockwise rotation. According to the input signal, the circular distributor gives the output signals A, B, C, representing the energizing state of coil windings of the three phases, respectively. The three signals are amplified by the power amplifier(功率放大器) to drive the motor to run. Some hardware distributor can also set the power sequence, so that the output can be selected from single-powered 3-step mode or double-powered 3-/6-step modes.

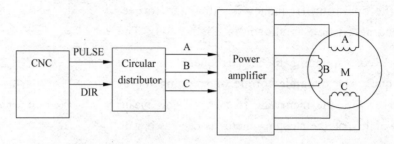

Figure 3-10　Hardware circular distributor connection

The software circular distribution is realized by the program in the microcontroller (微控制器程序) of the CNC or the driver. The output port of the microcontroller unit (MCU) directly gives the signal of each phase, which simplifies the hardware and improves the flexibility. The software circular distribution is usually achieved by the look-up-table method. As shown in Figure 3-11, the pin 0—2 of the MCU parallel output port PA is used for output the on/off signals for the A, B, C windings of the stepper motor. The signals are isolated by the photocouplers and sent to the power amplifier to drive the motor. Assuming that the bits 0—2 of the parallel output register of the MCU are correspond to the output state(输出状态) of the PA0—PA2 pins, the circular distribution table designed according to the three-phase 6-step mode is shown in Table 3-1. Save the control word in the look-up table in the memory of MCU. In the forward rotation, the address of the look-up table is increased by one in the circular mode (i.e. if the last address is 06H, the new address will be 00H), find the corresponding control word and write to the output register of PA port to complete a state switching. In the reversal rotation, the address is decreased by one (if the last address is 00H, the new address will be 06H). Therefore, the bidirectional rotation can be realized.

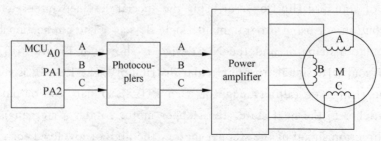

Figure 3-11　MCU-based software circular distribution circuit

Table 3-1 Software circular distribution table for the MCU

Direction		Energized winding	Output state	Address (hexadecimal)	Parallel output register control word (hexadecimal)
Forward	Reversal		CBA		
↓	↑	A	001	00H	01H
		AB	011	01H	03H
		B	010	02H	02H
		BC	110	03H	06H
		C	100	04H	04H
		CA	101	06H	05H

2) Power amplifier

The voltage of the control signal(控制信号 kòng zhì xìn hào) generated by the circular distribution circuit is generally only 3—5 V and the current is only several mA, thus it cannot directly drive the stepper motor. The control signal must be amplified before driving each phase. The semiconductor components(半导体元件 bàn dǎo tǐ yuán jiàn) that be used include gate-turn-off (GTO) thyristor, giant transistor (GTR) and power MOSFET, etc.

(1) Single-voltage power amplifier circuit

Figure 3-12 is the basic diagram of the single-voltage power amplifier to drive a single-phase coil winding of a stepper motor. V_{IN} is the phase coil control pulse after circular distribution, U is the motor power supply, L is the inductance of the motor winding, R_L is the winding resistance, R_1 is the current limiting resistor, VT is a power transistor. The VT operates in the switching mode, when V_{IN} is high, the VT will be turned on and the winding is energized. When V_{IN} is low, the VT will be turned off. VD is a freewheeling diode, which works along with R_2 to release the counter electromotive force (EMF) of L during turning off VT in order to protect VT. The current limiting resistor R_1 determines the time constant of the circuit. If the R_1 is too large, the winding current is decreasing and the rising time is increasing. A capacitor can be connected in parallel(并联 bìng lián) to R_1 to improve the high-frequency performance, but this will affect the low-frequency performance. Because R_1 needs to consume a certain amount of energy while working, the efficiency of this circuit is very low and it is only applied in some small power motors.

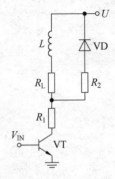

Figure 3-12 Single-voltage power amplifier circuit

(2) High-and low-voltage power amplifier circuits

The principle of high-and low-voltage power amplification is, using high voltage in the starting phase of energizing the coil winding, to make the current fast increased, then switch to the low-voltage power supply to maintain the current in the winding. The circuit of this type of the power amplifier is shown in Figure 3-13. The amplifier requires two power supplies to drive the motor, where U_H is a high-voltage power supply and U_L is a low-voltage power supply. V_H and V_L are used to control the on/off state of VT1 and VT2, respectively. When energizing the coil winding, V_H and V_L become high level simultaneously. Due to the effect of VD1, at this time only the high-voltage supply U_H is applied to energize the winding and the winding current increases fast. After reaching the specified value, V_H becomes low level but V_L is still in high level, thus the motor power supply will be switched to the low-voltage U_L. The winding current is maintained until V_L becomes low, and the power supply is switched off(断开). No matter how the frequency of operation, each energizing action for the winding is carried out in the above sequence, so we must ensure that the high-level time of V_H is shorter than that of V_L. The high-and low-voltage power supply method reduces the rising time of winding current, increases the working frequency of the motor and also increases the output torque at high frequency. Besides, the current of the current-limiting resistor R_1 is reduced by using the low voltage(低电压) to maintain the winding current. The disadvantage of this method is that, when switching the high voltage and low voltage, the top of the current waveform is concave, which affects the stability of the operation.

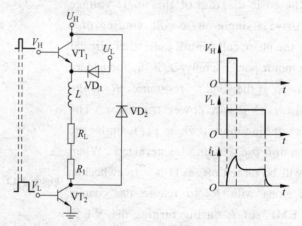

Figure 3-13　High-and low-voltage power amplifying circuits

(3) Chopper power amplifier circuit

The idea of chopper drive is to control the switching state of the power transistor according to current feedback during energizing signal of the winding, so that the winding current is maintained near the reference value(参考值), and the circuit is shown in Figure 3-14. In this circuit, A1 is an analog comparator, R_1 and R_2 work as voltage dividers and V_{REF} is the reference current signal that connected to the noninverted input of A1. R_3 is the current sampling resistance (the resistance value is very small), the voltage signal is generated by sampling the winding current and connected to the inverted input of A1. Therefore, when the reference current is higher than the armature current(电枢电流), the output of A1 is high; otherwise, the output of A1 is low. A2 is the logic AND gate, when the winding energizing command signal V_{IN} is "0", the output of A2 will be 0, VT will cut-off and the winding will not be energized; When V_{IN} is "1", the output of A2 will be the same state of the output of the comparator. Since V_{REF} is constant, when V_{IN} is "1", the switching state of VT will be determined by the feedback current, that is, when the feedback current is lower than the reference value, VT will be turned on to make the current rising; when the feedback current is higher than the reference value, VT will be turned off and the current will fall. During the time when the energizing signal V_{IN} is valid, fast turning on and off VT according to the current feedback can make the winding current fluctuate(波动) around the reference value, thus realize the constant current control and ensure a constant torque within a large frequency range. In this control mode, the value of R_3 is very small and the power dissipation is also small, therefore the efficiency is very high. Meanwhile, the circuit time constant is also very small, which guarantees a fast response. The disadvantage of this type of driving circuit is that the electromagnetic noise is relatively large.

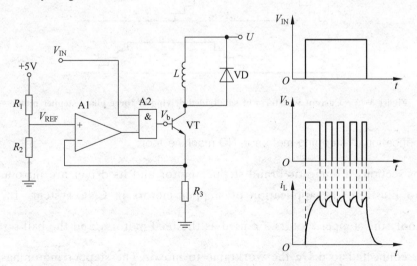

Figure 3-14 Chopper amplifier circuit

(4) Subdivision driving circuit

For the above-mentioned driving circuit, each phase winding has only two states of powering, i.e. ON and OFF, and the number of the directions of <u>winding magnetic field</u>

(绕组磁场) is completely determined by the combination of the ON/OFF state of all the phases in every step. If the current value of each phase is further subdivided to let the current rising/falling in a stair manner, then there will be more states can be combined to increase the number of steps and the available magnetic field directions, thereby reducing the step angle and improving the <u>precision</u>(精度) of the servo system. Figure 3-15 shows the waveform of each phase current of the three-phase stepping motor subdivided by 4. It can be seen that in each state there is/are one/two phase(s) energized, which is similar to the 6-step driving mode of the three-phase motor. However, the current of each phase is gradually increased or decreased according to the "stair" waveform, the number of energizing states is increased to 24, and the step angle is reduced to 1/4 of that in the 6-step mode. The current subdivision can be achieved using the above-mentioned <u>chopper power</u>(斩波功率) amplification with the reference voltage given by a D/A converter connected to the microcontroller. Besides, the subdivision can be also achieved by the pulse-width modulation (PWM) technology. At present, many manufacturers of stepper motor drivers support subdivision-driving technology and the maximum number of subdivision can be hundreds of steps.

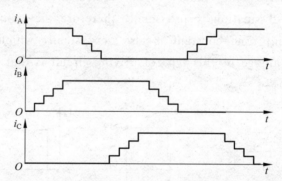

Figure 3-15　Current waveform of subdivided driving of three-phase stepper motor

5. Application of stepping motors in NC machine tools

In this section, a domestic brand stepper motor and its driver are introduced as an example to illustrate the application of stepper motors in CNC system. In the CNC machine tool, the stepper motors are used as the feed motors, and the <u>ball screws</u>(滚珠丝杠) are connected to drive the worktable to move. The stepper motor has 2 phases and the driver works in the chopper amplification mode with subdivision function. The operation panel of the driver is shown in Figure 3-16, which includes LED indicators, control signal terminals, an 8-bit DIP switch, power input terminals and motor current output terminals（端子 duān zǐ）, each part is shown in Table 3-2. In the DIP switch, SW1—SW3 are used to set the output current, the corresponding values are listed in Table 3-3; SW4 is used to set the static current (when motor is not moving), when SW4 is ON it will be the full driving current, when SW4 is OFF it will be the half driving current (to save energy and generate less heat, but the torque is smaller); SW5—SW7 are used to set the subdivision, i. e. set the number of steps in a revolution, see Table 3-4. It can be seen that the driver can achieve up to 25,000 steps per revolution by subdivision.

Figure 3-17 shows some different wiring modes for the driver and stepper motor. Figure 3-17(a) is the basic wiring to connect with the 2-phase 4-wire motor; Figure 3-17(b) is the serial wiring method for the 8-wire stepper, which has little vibration（振动 zhèn dòng） in the low speed; Figure 3-17(c) is the parallel wiring method for the 8-wire motor, which increases the driving current and improves torque in high speed;

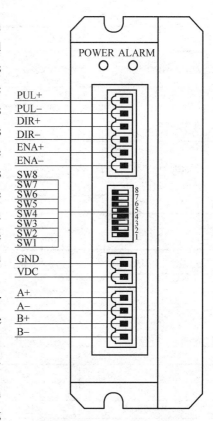

Figure 3-16 Stepper motor driver

Figure 3-17(d) is the low-speed mode wiring for a center-tapped 6-wire stepper motor, which can provide the maximum torque of the motor; and Figure 3-17(e) is the high-speed mode wiring for the 6-wire motor, which has a smooth motion in high speed.

Table 3-2 Function description for the stepper motor driver interface

Function	Symbol	Description
Indicator LED	POWER	Power indicator. Turn on when the power supply is applied
	ALARM	Alarm indicator, which indicates over-current, overheat, overvoltage and so on
Control signal terminals	PUL+	Differential input terminals for the pulse signal. The number of pulses corresponds to the rotation angle and the frequency of the pulses corresponds to the rotation speed
	PUL−	
	DIR+	Differential input terminals for the direction signal. Used to determine the rotation direction of the motor
	DIR−	
	ENA+	Motor-enable signal input. When the signal is valid, the motor will be energized; Otherwise, the motor will be released
	ENA−	

continuous

Function	Symbol	Description
DIP switch	SW1-SW8	For the setting of current, subdivision and other functions
Power input terminals	GND	The ground of the DC power supply
	VDC	The positive input of the DC power supply. Supply voltage: 20—50 V
Motor current output terminals	A+	Motor A-phase
	A−	
	B+	Motor B-phase
	B−	

Table 3-3 Current settings for the step motor

Peak current/A	Average current/A	SW1	SW2	SW3
1.00	0.71	ON	ON	ON
1.50	1.06	OFF	ON	ON
2.00	1.42	ON	OFF	ON
2.50	1.77	OFF	OFF	ON
3.00	2.12	ON	ON	OFF
3.50	2.48	OFF	ON	OFF
4.00	2.82	ON	OFF	OFF
4.50	3.20	OFF	OFF	OFF

Table 3-4 Subdivision settings for the stepper motor

Steps/Rev	SW5	SW6	SW7	SW8
400	OFF	ON	ON	ON
800	ON	OFF	ON	ON
1,600	OFF	OFF	ON	ON
3,200	ON	ON	OFF	ON
6,400	OFF	ON	OFF	ON
12,800	ON	OFF	OFF	ON
25,600	OFF	OFF	OFF	ON
1,000	ON	ON	ON	OFF
2,000	OFF	ON	ON	OFF
4,000	ON	OFF	ON	OFF
5,000	OFF	OFF	ON	OFF
8,000	ON	ON	OFF	OFF
10,000	OFF	ON	OFF	OFF
20,000	ON	OFF	OFF	OFF
25,000	OFF	OFF	OFF	OFF

The control signals of the driver are <u>electrically isolated</u>(电隔离) by <u>photoelectric couplers</u>(光电耦合器), which improve the noise-resistance ability and can support 3

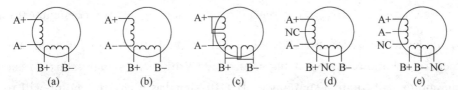

Figure 3-17 Wiring modes for the stepper motor and driver
(a) 4-wire motor; (b) 8-wire motor series mode; (c) 8-wire motor parallel mode;
(d) 6-wire motor low-speed mode; (e) 6-wire motor high-speed mode

different connection types. In Figure 3-18, (a) is the common anode connection mode, where the output signals of the CNC system use NPN-type transistors; Figure 3-18(b) is the common cathode(阴极)connection mode, where the outputs of the CNC system use PNP-type transistors; and Figure 3-18(c), the CNC employs differential outputs, in which each signal is transmitted by a twisted-pair cable to improve the ability at suppressing the common mode noise, which can realize long-distance signal transmission.

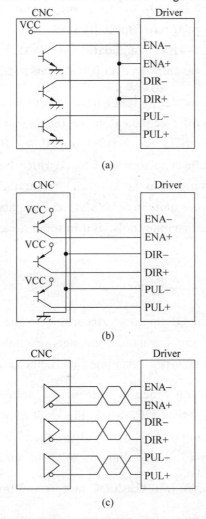

Figure 3-18 Wiring of the CNC system and the driver

The CNCs and the drivers use the pulse/direction signals for communication, and the signal waveforms(波形) are shown in Figure 3-19. If ENA signal is valid, the motor will be energized and ready for being controlled. With ENA signal valid, on each PUL pulse the motor will rotate a step angle. If DIR signal is high, the motor will rotate in the forward direction; if DIR signal is low, the motor will rotate in the reversed direction.

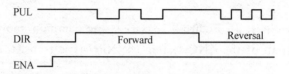

Figure 3-19 Control signal waveform

3.2.2 DC servo motors

Direct current (DC) motors were the first type of motors widely used, and also the first type to realize speed regulation. Especially since 1970s, with the appearance of the power transistor and pulse-width modulation (PWM) technology, the control performance of DC motors was improved so it had been widely used in the field of CNC machine tools. Since DC motors(直流电机) have the advantages of high precision, high efficiency, fast response, hard mechanical characteristics, etc., they had been widely applied in the speed-control field for a very long time. However, DC motors have complex structure and are subject to wear and tear, which has high requirement for the operating environment and needs to be carefully maintained, so in recent years the permanent-magnet synchronous motors (PMSMs) have gradually replaced the traditional DC servo motors in the CNC machine tools. But in some situations, the DC servo motors (直流伺服电机) are still used.

1. Classification of DC motors

There are many types of DC motors. According to the excitation modes, the DC motor can be divided into the permanent-magnet, excitation, shunt-excited, series-excited and combined-excited types. Among them, the permanent-magnet type uses ferrite, Al-Ni-Co, Nd-Fe-B and other permanent magnetic materials(永磁材料) to establish the excitation magnetic field, while the excitation type needs the excitation windings. According to the moment of inertia of the rotor, it can be divided into small-, medium-and large-inertia types. According to ways to realize the switch of current direction, they can be classified into brush DC motors and brushless DC motors(无刷直流电机). In CNC machine tools, the feed servo system often employs small-inertia and wide-speed-regulation capability DC servo motors, while the spindle servo system employs the DC spindle motors.

1) Small-inertia DC servo motor

The rotor of the small-inertia DC servo motor is long in length and small in diameter, therefore the moment of inertia is very small, only about 1/10 of the ordinary DC motor. The small-inertia DC servo motor has the advantages of fast response, good dynamic performance(动态性能) and can run smoothly at low speed. However, the overload capacity of the small inertia DC servo motor is weak, and the inertia of the rotor is less than the inertia of the moving parts of the machine tool, thus it is usually necessary to use reduction gears for being connected to the lead screw, which increases the transmission error(传动误差). Small-inertia DC servo motors had been widely used in the early CNC machine tools.

2) Large-inertia DC servo motor with wide speed regulation capability

The large-inertia and wide-speed-regulation capability DC servo motors were developed based on the small-inertia DC servo motors. It keeps the large torque/inertia ratio and improves the dynamic response characteristics by increasing the torque. Because of its wide speed range and large inertia, it can be directly connected to the lead screw and does not need intermediate transmission parts, which simplifies the transmission chain and improves the response performance and the stability of each servo axis. Large-inertia DC motor can usually run smoothly at low speed of 1 r/min or 0.1 r/min. In addition, the structure of the large-inertia motors provides a relatively large heat capacity(热容量), which results in large overload capacity and allows running under the condition of large current for dozens of minutes. Therefore, it had been widely used in CNC machine tools since 1970s.

The large-inertia and wide-speed-regulation capability DC servo motors can be divided into electric-excited and permanent-magnet types. The strength of the excited magnetic field of electric-excited type is easy to adjust, and to set the compensation winding and the commutating pole, thus the commutation performance(换向性能) is good and can achieve constant-torque-speed regulation in a wide speed range. The permanent-magnet motors generally do not have the compensation winding and the commutating pole, therefore the commutation performance is limited. But they do not need excitation power for the magnetic field, so the efficiency is high and the output torque is large at low speed. In addition, this type of motors has a compact structure(紧凑结构) and low temperature rising. Besides, the performance of permanent magnet materials is continuously improving, so their applications are more and more.

3) Brushless DC servo motor

The brushless DC motor eliminates the mechanical brush and the mechanical commutator, and uses a position sensor and an inverter to realize commutation, so it has higher reliability and longer service life. In fact, the brushless DC servo motor is a kind of AC synchronous motors(交流同步电机). Its speed-regulation performance can reach the same level as that of the DC servo motor. Besides it provides high reliability, it has drawn close attention to the engineers in recent years.

2. The working principle of DC servo motors

The working principle of DC servo motors is the same as that of the common DC motors. The basic structure of the DC motor comprises a stator (magnetic pole), a rotor (armature), an electric brush and a commutator. In this section, we study the characteristics of the excitation-type DC servo motor. As shown in Figure 3-20(a), the excitation winding on the left generates(产生) a magnetic field. At this time, the armature of the rotor in the right side is energized with the direct current, and the rotor rotates(转子旋转) under the effect of the electromagnetic torque, and the inductive electromotive force is made by cutting the magnetic-induction lines. Its equivalent circuit is shown in Figure 3-20(b). When the circuit is balanced, the circuit voltage(电压) equation can be established as:

$$U_a = E_a + I_a R_a \tag{3.2}$$

where U_a is the terminal voltage of the armature circuit, E_a is the electromotive force of the armature winding, I_a is the armature current, and R_a is the total resistance of the armature circuit.

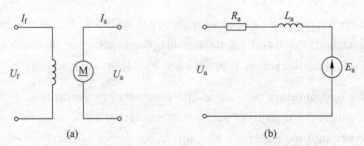

Figure 3-20　Working principle of the excitation-type DC motor
(a) Working circuit; (b) Equivalent circuit

Assuming that the magnetic flux Φ, generated by the excitation circuit, is constant, and the induced electromotive force of the armature winding is proportional to the rotational speed, then

$$E_a = C_E \Phi n \tag{3.3}$$

where C_E is the electromotive force constant (the inductive electromotive force per speed unit), and n is the motor speed.

The electromagnetic torque of the motor is:
$$T_m = C_T \Phi I_a \tag{3.4}$$
where C_T is the electromagnetic torque constant (electromagnetic torque generated by a unit current), T_m is the electromagnetic torque(电磁转矩) of the motor.

The equation of the mechanical characteristic of the excitation-type DC servo motor can be found by solving (3.2)—(3.4) as follows:
$$n = \frac{U_a}{C_E \Phi} - \frac{R_a}{C_E C_T \Phi^2} T_m = n_0 - \frac{R_a}{C_E C_T \Phi^2} T_m \tag{3.5}$$
where
$$n_0 = \frac{U_a}{C_E \Phi} \tag{3.6}$$
is the ideal speed of the DC servo motor with no load.

The mechanical characteristic of the DC servo motor refers to the function relationship between the motor speed and the torque under the condition of stable running, in which the electromagnetic torque is equal to the load. The speed difference, Δn, is the one between the actual speed and the ideal speed when the motor is loaded. The value of Δn represents the "hardness" of the mechanical characteristic(机械特性), the smaller the Δn, the harder the mechanical characteristic. According to the mechanical characteristic equation (3.5) of DC servo motor, there are three basic methods to adjust the speed.

1) Changing the armature voltage U_a

This method keeps a constant excitation flux Φ and changes the armature voltage to achieve speed regulation. The mechanical characteristic curve is shown in Figure 3-21(a). It can be seen that when the armature voltage is changed, the mechanical characteristics of the DC motor is a set of parallel lines, the slope of the characteristic curve(特性曲线) remains unchanged, and the mechanical characteristic is maintained. If the current feedback control is added based on this, the armature current is kept constant by adjusting the armature voltage, and the torque of the motor, T_m, can be kept to achieve constant torque control. Since this method has good mechanical characteristic, it is suitable to be applied to the low-speed range in the feed servo driver and spindle driver of the CNC machine tools.

2) Changing the excitation flux Φ

The mechanical characteristics of this method are shown in Figure 3-21(b). Changing the flux changes not only the ideal no-load speed, but also the slope of the mechanical characteristics, so that the mechanical(机械的) characteristic becomes soft. This method is not suitable for the feed servo driver. It can be used in the high-speed range of the spindle driver. It should also be noted that, for the permanent magnet DC servo motor it is impossible to change the excitation flux Φ to achieve speed regulation.

3) Changing the resistance of the armature circuit R_a.

In this method, the mechanical characteristic is very soft, and the efficiency is low, so it is difficult to achieve high-power stepless speed regulation (无级调速). Therefore, it is not suitable for the servo control of CNC machine tools.

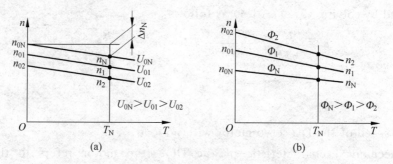

Figure 3-21 Mechanical characteristic of the DC servo motor
(a) Mechanical characteristic when changing the armature voltage;
(b) Mechanical characteristic when changing the magnetic flux

3. Driving and control of DC servo motors

The driving of DC servo motors in the CNC machine tools depends on power semiconductor devices. In the early age, the thyristor is widely used in the servo drivers, in which the voltage regulation is achieved by adjusting the operating angle of the thyristor by phase shifting. The thyristor has a large-rated current, strong overload capacity(过载能力) and simple-control circuit. However, the thyristor(晶闸管) servo system has low-control frequency, large-current fluctuation in the low-speed range and slow response. Later, with the development of power semiconductor technology, the gaint transistor (GTR), power metal-oxide semiconductor field-effect transistor (MOSFET), insulated gate bipolar transistor (IGBT) and other new devices are introduced, which enables the pulse-width modulation (PWM) technology to be easily applied in the servo system. They had overcome the disadvantages of the thyristor driver and greatly improved the performance of DC servo systems. Therefore, the PWM control system is now the main control method for the DC servo motors.

The basic principle of the DC servo motor speed control system based on PWM is shown in Figure 3-22. The power supply of the circuit is a constant DC voltage U_s, the voltage applied on the DC motor U_O can be controlled by turning on/off of the switch. In the actual control circuit(实际控制电路), the switch is realized by

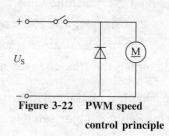

Figure 3-22 PWM speed control principle

a power semiconductor(半导体) device, and the on-off state can be controlled according to the switch signals of the servo driver. In the PWM control mode, the total time of the on/off cycle is a constant T, and the voltage applied on the motor terminals is a sequence of square wave, as shown in Figure 3-23. In this figure, the voltage is U_s when the switch is ON, and the voltage is zero when the switch is OFF. If the ON time in a cycle is t, the average voltage at the two ends of the armature winding of the DC motor is

$$\overline{U}_o = \frac{t \cdot U_s + (T - t) \cdot 0}{T} = \frac{t}{T}U_s = \alpha U_s \qquad (3.7)$$

where α is the duty cycle, $\alpha = \frac{t}{T}$, ranging in $0 \leqslant \alpha \leqslant 1$. Therefore, by changing the duty cycle, the average voltage between the two ends of the armature can be continuously changed between $0 - U_s$ to realize stepless speed regulation. As can be seen from Figure 3-23, in the PWM control, the armature voltage is in a pulse waveform. Due to the effect of the armature inductance(电枢电感), the current waveform is of the continuous "wave" type. Therefore, the PWM control will produce fluctuations in the current and torque. By increasing the frequency of PWM, the current fluctuation can be suppressed significantly, and the fluctuation of the torque and speed can be also reduced.

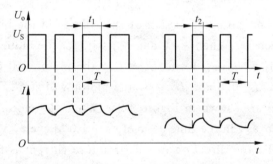

Figure 3-23 Voltage and current waveforms of the PWM DC motor

The motor in Figure 3-22 can rotate only in one direction, which called irreversible speed-regulation system. For the position control of the feed servomechanism of the CNC machine tool, the motor must be able to rotate in both directions. The power amplifier driving circuits of the reversible(可逆的) PWM system can be classified as T type and H-type, and the H-type circuit is applied much more. The H bridge power amplifier circuit is shown in Figure 3-24, where U_s is the power

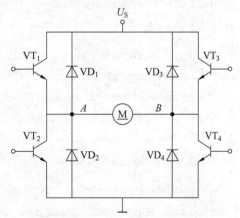

Figure 3-24 H bridge power amplifier circuit

supply, VT1—VT4 are power transistors, VD1—VD4 are freewheeling diodes. According to the different driving methods, the reversible PWM system has two driving modes: bipolar-(双极的) and unipolar-(单极的)driving, both of which use a same driving circuit. In the bipolar-driving mode, the armature voltage of the DC servo motor has positive and negative voltage states in a PWM cycle; While, in the unipolar-driving mode, in a PWM cycle the armature only bear one voltage state (positive voltage or negative voltages).

Take the bipolar reversible driving as an example, the 4 power transistors are divided into two control groups: group 1 is VT1 and VT4, group 2 is VT2 and VT3. In a PWM cycle, the transistors(晶体管) in a same group are synchronously turned on or off, and the ON/OFF states of the transistors in different groups are the opposite. If the PWM control signal given by the control is in high level, VT1 and VT4 will be turned ON while VT2 and VT3 will be turned OFF, and the armature winding will bear the positive(正的) voltage from A to B; If the PWM signal is switched to the low level, VT1 and VT4 will be turned OFF and VT2 and VT3 will be turned ON, the armature winding will bear a negative voltage from B to A, that is, the bipolar driving. In a PWM period, the average voltage of the armature is:

$$\bar{U}_\mathrm{o} = \frac{tU_\mathrm{s} - (T-t)U_\mathrm{s}}{T} = \left(2\frac{t}{T} - 1\right)U_\mathrm{s} = (2\alpha - 1)U_\mathrm{s} \qquad (3.8)$$

It can be seen that the magnitude and sign of the voltage in the bipolar-driving method depend on the duty cycle α. When $\alpha = 0$ and $\bar{U}_\mathrm{o} = -U_\mathrm{s}$, the motor will reversely rotate at the maximum speed; when $\alpha = 1$ and $\bar{U}_\mathrm{o} = U_\mathrm{s}$, the motor will forward rotate at the maximum speed; when $\alpha = 0.5$ and $\bar{U}_\mathrm{o} = 0$, the motor will stop. The armature voltage and current waveforms are shown in Figure 3-25. When the motor is rotating forward, as shown in Figure 3-25(a), in the time $0-t$ of each PWM cycle, VT1 and VT4 are ON, VT2 and VT3 are OFF, the direction of the armature current is from A to B and the current increases. In the time $t-T$, VT1 and VT4 are OFF, VT2 and VT3 are ON, the winding bears the reverse voltage. However, since the t is relatively larger in the forward working phase, thus the forward current is larger, the current direction in t-T remains unchanged, and the magnitude of the current(电流) decreases continuously. After the beginning of the next PWM cycle, the current will continue to increase. Therefore, in the forward rotation case, the current direction is from A to B and the magnitude of the current is fluctuating. The reverse case is similar to that in the forward case, as shown in Figure 3-25(b). As can be seen from Figure 3-25(c), at low speed and with light load, due to the effect of PWM switching, the current oscillates between the positive and negative directions(方向), which is helpful to eliminate the static friction of the load, so the smoothness at low speed is good. It can be seen that the magnitude and direction of the armature current can be controlled by changing the duty cycle, so as to realize the control over the torque and speed.

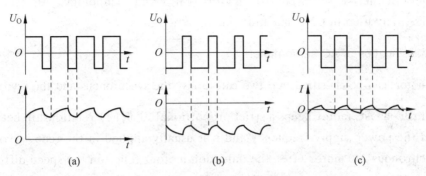

Figure 3-25 Voltage and current waveforms of H bridge bipolar reversible driving
(a) Forward; (b) Reversal; (c) Light load

In practical applications of the DC speed control system, to achieve better performance, the double-closed-loop feedback of the current and speed PWM control system is employed. The block diagram is shown in Figure 3-26. Double-closed-loop feedback control can improve the control precision and dynamic response characteristic of DC speed-regulation system, and achieve good control quality. The control and driving system includes a speed regulator, a current regulator, a PWM controller, a driver for the base of transistor, a power amplifier, and the current and speed feedback
sù dù fǎn kuì
(速度反馈). Among them, the speed regulator, the current regulator and the PWM controller can be realized by microcontrollers or other embedded microprocessors. In particular, nowadays many microcontrollers have the hardware PWM control function, which simplifies the controller development as well as achieves full digital control.

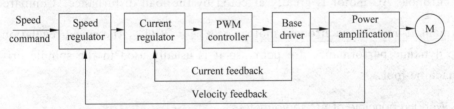

Figure 3-26 Double-closed-loop PWM DC speed regulating system

3.2.3 AC servo motors

DC servo motors have good performance and are simple in control. However, due to the easy abrasion of the brush and commutator, it is prone to produce the spark when
zhì zào gōng yì
commutation. The structure and manufacturing process(制造工艺) of DC servo motors are complex, and the cost is high, therefore the application is limited. Since the AC motor eliminates the brush and the structure is simple, the reliability is high and the output power is large. The control of AC servo motors is relatively complex. With the

development of the AC servo control theory(理论) and technology, AC servo motors have been widely used in CNC machine tools.

1. Classification of AC servo motors

AC motors can be classified into two categories: the synchronous and the asynchronous. The synchronous AC motor keeps a strict proportional(比例的) relationship between the speed and the power supply frequency and it is usually applied in the feed servo system. The asynchronous AC motor (i.e. the induction motor) relies on the speed difference to produce electromagnetic torque, so the motor speed is slower than the synchronized one. But its structure is relatively simpler and the cost is lower, and it is usually used in the spindle driving system.

The synchronous AC motors, according to the method to establish the air gap(气隙) magnetic field, can be divided into the electromagnetic and the non electromagnetic. The latter can be divided further into the permanent-magnet, the reactive and the hysteresis types. The rotor of PMSM uses permanent-magnet material and the stator has symmetrical multiphase sinusoidal-distributed windings, which has simple structure, good regulation performance, high reliability, high efficiency and small size, thus suitable for speed-regulation system and high-precision position servo system.

Asynchronous AC motors can be divided into rotor-winding-induction motors, cage-type-induction motors, etc. They do not need to use permanent magnets, so the weight is light in the large-power-capacity applications and its cost is low. However, the speed of asynchronous AC motor is greatly affected by the load disturbance. Compared with the permanent-magnet synchronous motor(永磁同步电机), the speed-regulation range and the dynamic performance are poor, so it is usually used in the spindle driving of CNC machine tools.

2. Working principle of AC servo motors

1) Permanent-magnet synchronous motor (PMSM)

The structure of the PMSM servo motor is shown in Figure 3-27, including the stator, the rotor and detecting elements(检测元件). The stator is provided with slots, wherein the three-phase windings or multi-phase symmetrical windings are distributed. The rotor part is provided with a permanent magnet to generate the magnetic field, which is made of high-performance rare-earth permanent-magnetic materials. Figure 3-28 is a typical protruding rotor structure(突出转子结构), which have tile-shaped permanent magnets on the surface of the rotor. In addition, the installation of the rotor magnet also includes plug-in, radial, tangential and mixed types, different installation methods have impact on the magnetic circuit and parameters of the motor. The detecting element is installed on the motor shaft to detect the rotor position, for which the pulse encoders or rotary transformers can be employed.

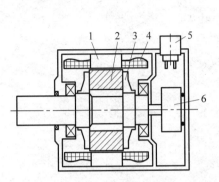 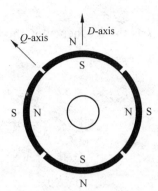

Figure 3-27　Structure of the PMSM servo motors

1—stator; 2—rotor; 3—plate; 4—three-phase winding of stator;
5—wiring box; 6—pulse encoder

Figure 3-28　Rotor structure of PMSM

If the balanced three-phase AC is applied to the stator of the PMSM, a <u>space rotating</u>
　　　　　　　kōng jiān xuán zhuǎn cí chǎng
<u>magnetic field</u>(空 间 旋 转 磁 场)with an angular velocity of ω can be produced. This magnetic field interacts with the rotor magnetic field so that the rotor rotates with the synchronized angular velocity ω. As shown in Figure 3-29(a), the three-phase balanced current i_A, i_B and i_C are applied to the space-static three-phase A, B, C in the stator windings to produce a composite rotary magnetic flux Φ, which is sinusoidal-distributed in space, rotating with the angular velocity ω. Such a rotating flux can be also produced by applying balanced two-phase current i_α, i_β with 90° phase shift to two space-static perpendicular windings. If the magnitude and the angular speed of this magnetic flux is equal to the three-phase winding flux Φ, the two-phase windings will equivalent to the three-phase windings. In Figure 3-29(c), two perpendicular windings with the same number of turns d and q are energized by the direct current i_d and i_q, meanwhile, the two DC windings rotate with the angular speed ω, thus the rotating magnetic flux produced by them also rotates with the angular speed ω, which is also equivalent to the rotating magnetic field of the stationary three-phase windings. In practice, the axis of the magnetic pole of the rotor of the PMSM is defined as d-axis. By applying the above-
　　　　　　　　　　shī liàng biàn huàn
mentioned <u>vector transform</u>(矢 量 变 换), the rotating magnetic field generated by the space-static three-phase windings is transformed to the equivalent magnetic field of
　　　　　　　　　　　　　　　děng xiào de
rotating DC windings, and the <u>equivalent</u>(等 效 的) DC current i_d and i_q are used as control variables, the decoupling control of PMSM can be realized, just like the control of a excited-type DC motor. For example, let $i_d = 0$ and only change i_q, is equivalent to

change armature voltage of the DC motor to achieve speed regulation; if change i_d, flux-weakening control can be achieved similar to change the excitation field in the DC motor. For the motor with p-pairs of poles, the synchronous speed n is:

$$n = 60f/p \qquad (3.9)$$

where f is the frequency of the AC in the stator.

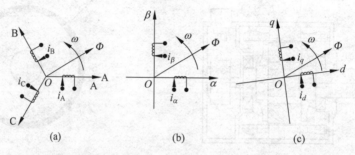

Figure 3-29　Equivalent transformation model of PMSM windings

The starting process of PMSM is more complex than that of the DC motor, since the large inertia moment of the rotor, and the difference in angular speeds of the magnetic fields between the stator and the rotor is too large. If starting directly at high speed, the average electromagnetic torque applied to the rotor at the start time is close to zero, so it is difficult to move. Therefore in the starting process of PMSM, the motor should be gradually accelerated(加速) from low speed to high speed.

2) Asynchronous motors

The AC asynchronous motors used in the spindles of CNC machine tools are very different from the PMSMs used in the feed servo systems. The asynchronous motors are based on the principle of induction motors(感应电动机). Due to the fact that the spindle of the machine tool needs high power but no special requirements for the continuous position follow, from the view of cost, the rotor part adopts a specially designed cage structure(笼型结构) rather than the permanent-magnet structure. The cylinder rotor is provided with evenly distributed conducting bars, and both ends of the conducting bars are connected together by metal rings to form a cage structure. In order to increase the output power and reduce the volume, the iron core of the stator is directly cooled in the air, therefore eliminates the motor case. The shape of the spindle motor is mostly polygon, rather than the common circle-shaped asynchronous motor(异步电动机). A rotary encoder is installed at the tail of the motor shaft for position detection. In recent years, the built-in-type spindle motor, which integrates the structure parts of the spindle of the machine tool and the spindle motor, has been developed. This type of spindles is driven directly by the motor and has no transmission mechanisms, also known as the motorized spindle. The built-in type spindle motor has the advantages of compact structure, small inertia, high precision and stability, and can realize the high-speed control of the spindle.

When the three-phase alternating current is applied to the stator in an asynchronous AC motor, a rotating magnetic field can be generated. The magnetic field cuts the rotor conductor(转子导体)and generates induced current in the rotor, which interacts with the magnetic field of the stator to produce an electromagnetic torque(电磁转矩) that drives the rotor to rotate. It can be seen from the above principle of asynchronous AC motors that only when the rotor and stator have different speeds, the magnetic field can cut the conductor(导体) to produce electromagnetic torque. For the asynchronous motor with p-pair of poles, the speed n_r is

$$n_r = n_s(1 - s) = \frac{60f}{p}(1 - s) \qquad (3.10)$$

where f is the frequency of the AC in the stator, n_s is the synchronous speed, s is slip ratio, $s = (n_s - n_r)/n_s$. If the load is increased, the slip will be increased. In practical application, the speed regulation of AC asynchronous motor is mainly based on the variable-voltage variable-frequency (VVVF) method and vector control method. The VVVF method is relatively simple, but in the process of speed regulation the mechanical characteristics may become poor. The vector control method can be used to control the speed of induction motor by means of the coordinate transformation(坐标变换), which lead to a similar speed-regulation method as used in the DC motors.

3. Speed regulation characteristics of AC servo motors

When the AC motor operates at the rated voltage and frequency, the stator iron core is close to saturation, and the temperature rising is in the rated range. In the variable-frequency speed regulation of AC motor, since it is constrained by the current, voltage, magnetic circuit(磁路), temperature rising and other aspects, the speed control scheme should be reasonably designed to exploit the performance of AC motors. As shown in Figure 3-30, when apply speed regulation below the rated frequency f_N of the motor, since it is limited by the stator core saturation, the maximum output of the electromagnetic torque is the rated torque T_e, known as constant torque speed regulation. In such a speed regulation, the mechanical characteristic is hard. In order to maintain a constant flux, the stator voltage is proportional to the frequency, as

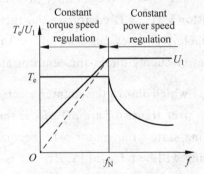

Figure 3-30 Characteristic of variable-frequency speed regulation of AC motor

shown in the dashed line(虚线) in the Figure. In practical applications, it is necessary to raise the stator voltage to compensate(补偿) the stator voltage drop at the low frequency, as shown in the solid line(实线) in the Figure. When the motor speed is above the rated frequency f_N, it is limited by the rated voltage, therefore cannot continue to increase the speed by increasing the stator voltage. In this situation, the flux will decrease with the increase of the frequency, the maximum torque output will decrease, and the motor characteristic will become soft. However, due to the rising speed, the power of the spindle is almost unchanged, known as constant-power speed regulation(恒功率调速).

4. Driving of AC servo motors

In the CNC machine tool, the AC servo motor driving device for the feed system is called the servo driver. In addition, some spindles in the machine tools also have position servo function and connected with a servo driver, which can be used as a feed axis.

In the driving device of the AC motor, the adjustable-frequency/voltage three-phase AC output is required to realize variable-frequency speed regulation. In order to convert the fixed-frequency AC input from the power grid to the adjustable-frequency AC output, the power semiconductor devices are used to realize the three-phase inverter circuit. Figure 3-31 is the main circuit of the AC-DC-AC inverter. In this Figure, VD_7—VD_{12} comprises the three-phase rectifier bridge. The capacitor C_d filters the rectified voltage and ensures the stability of the DC voltage when the load changes. The inverter transistor VT_1—VT_6 comprises the three-phase inverter bridge, which can generate the adjustable-frequency pulse-waveform three-phase (U, V, W) AC power output from the DC power supply under the appropriate switching signal. VD_1—VD_6 are freewheeling diodes, which absorb the counter-electromotive force in the motor coil(电机线圈) when the inverter transistor are off. R_1 is the braking resistor. When the AC motor is in the breaking state, the motor will become a generator and the regenerated current by breaking will be released by R_1. R_2 is a current limiting resistor. When the inverter is just energized, the current impact is very large, and the current is limited by R_2. After a certain time, S1 is closed to short the circuit current limiting resistor R_2.

With U-phase as an example, assume that the voltage of DC side is U_d, and let the midpoint voltage of the output of the inverted three-phase AC power be zero. If VT_1 is ON and VT_4 is OFF, the U-phase will output a $+U_d/2$ voltage relative to the midpoint; If VT_1 is OFF and VT_4 is ON, the U-phase will output a $-U_d/2$ voltage relative to the midpoint. As a result, the transistors on the same bridge arm are alternately switched

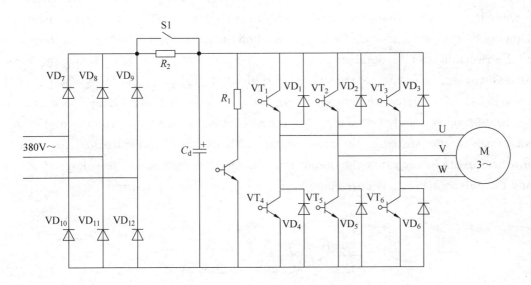

Figure 3-31 Main circuit of the AC-DC-AC inverter

ON, which can generate a pulse AC output. Varying the pulse width and frequency, the sine-wave pulse-width modulation (SPWM) can be achieved. In order to prevent short circuit(短 路 duǎn lù), VT_1 and VT_4 cannot be switched ON at the same time, thus it is necessary to make sure that VT_1 and VT_4 are OFF before switch ON one of VT_1 and VT_4.

The output current of the three-phase inverter is equivalent to the three-phase sinusoidal AC. According to the control theory, the effects of the pulses with equal impulse (冲 量 chōng liàng) and different shapes are same when applying on an inertia element. Therefore, in the AC motor control, a series of rectangular pulses can be used instead of the sine wave. Figure 3-32 is the equivalent principle of the rectangular pulse and the sine waveform. The area of each pulse is equal to the area of the corresponding sine waveform, so they have the same effect.

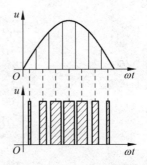

Figure 3-32 Equivalent principle of the rectangular pulse and the sine wave

In practical applications, the sine-wave PWM (SPWM) technology can be used to control the inverter bridge. The SPWM signal can be obtained by "modulating" the sine wave using the triangular "carrier wave". As shown in Figure 3-33(a), the three-phase SPWM sharing an isosceles triangle(等腰三角形 děng yāo sān jiǎo xíng) wave as the carrier to modulate the three-phase sine wave with phase differences of 120°. The principle of modulation is to compare the amplitude of the carrier and the modulated sine wave, when the sine wave amplitude is greater than the carrier amplitude(载波振幅), the output is high; Otherwise, the output is low. Using this method, three bipolar SPWM waveforms with 120° phase-shift can be obtained, as shown in Figure 3-33(b)—(d). Using the three SPWM waveforms to control the switching state of the bridge in the U, V, W phase of the AC-DC-AC inverter, the output voltage of each phase will also change according to the SPWM signal, and the voltage amplitude is $\pm u_d/2$. The line voltage of the AC motor can be composited by the phase voltage. For example, in Figure 3-33 (e), $u_{UV} = u_U - u_V$. It can be seen that the output line voltage is unipolar and the voltage is $\pm u_d$, and the pulse waveform is distributed according to the sine waveform.

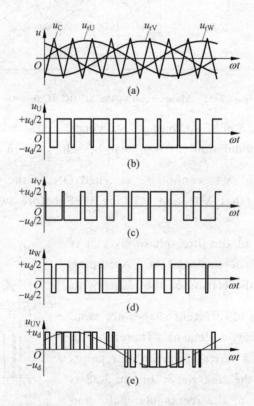

Figure 3-33　Waveform of the three-phase bipolar SPWM inverter circuit

The modulation of SPWM can be realized either by the analog circuit or by the method of computing. In practical applications(实际应用), the carrier frequency is much higher than the sine-wave frequency and the pulse period is very short, therefore by the filter effect of the inductance of the motor, the winding-current waveform is similar to the sine wave. In the modulation, the frequency and amplitude of the sine wave(正弦波) can be changed, and the frequency and the voltage of the equivalent sine-wave output of the inverter(逆变器) can be adjusted to realize the variable-frequency/voltage control of the AC motor.

5. Application of AC servo motors

The following example shows the application of the PMSM servo used as the feed motor of the CNC machine tool. The PMSM is connected with the ball screw, the screw nut drives the worktable to move, and the rotation of the motor is converted into the linear displacement of the worktable. A rotary encoder(编码器 biān mǎ qì) is attached to the servo motor, which measures the actual rotation angle of the motor for comparing to the command to achieve close-loop control. The servo motor should be driven by a suitable servo driver. The front panel of a domestic brand PMSM driver is shown in Figure 3-34, which has a LED display, operation buttons, the main power input terminals (L1, L2, L3), control power connection terminals (L1C, L2C), external regenerative resistor terminals (P, B1, B2), servo motor connection terminals (U, V, W), the earth screws, the host I/O signal connector (X1) and encoder connector (X2). The LED display and operation buttons can show the status of the motor and driver, display/save/modify the driver parameters, run test motion and show alarm history. The encoder connector is connected to the incremental encoder of the servo motor. The host I/O signal connector is used to connect with the CNC. The functions of other terminals are listed in Table 3-5.

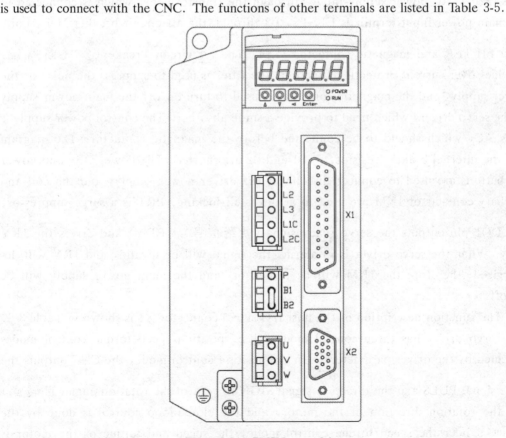

Figure 3-34 PMSM servo driver

Table 3-5 Functional description of the terminals of PMSM servo driver

Function	Symbol	Description
Main power input terminals	L1, L2, L3	Three-phase 220 V, 50 Hz AC power input. The 220 V voltage can be obtained from the 380 V AC power by a step-down transformer.
Control power connection terminals	L1C, L2C	Single-phase 220 V, 50 Hz AC power input.
External regenerative resistor terminals	P, B1, B2	When using the internal resistance of the driver, short B1 and B2 while disconnect P. If the external braking resistor is needed, wire the external braking resistor across P and B1, and B2 should be disconnected.
Servo motor connection terminals	U	Connect to the U-phase of PMSM
	V	Connect to the V-phase of PMSM
	W	Connect to the W-phase of PMSM
Earth screws	⏚	Connect the protective earth of PMSM and driver to the earth-ground.

The main circuit of the PMSM servo is shown in Figure 3-35. T, S, R are the three-phase 220 V AC power supply input (through the transformer), which is connected to the main power input terminals L1, L2, L3 through the air circuit breaker 1QF, noise filter FILTER and magnetic contactor 1KM. The air circuit breaker (空气断路器 kōng qì duàn lù qì) provides over current protection function, the filter is used to suppress the noise of the power supply, and the magnetic contactor is used to turn on/off the main power supply of the servo driver, which need to provide a surge absorber. The control power supply is 220V AC, which should to be connected before the magnetic contactor. The diagram uses the internal braking resistor, so B1 and B2 are shorted. The Power OFF and Power ON buttons are used to connect/disconnect the driver power supply, and the coil and auxiliary contacts of 1KM are used to achieve self-locking. PRT is a surge suppressor.

The DO1 pin outputs the servo ready signal (就绪信号 jiù xù xìn hào) (RDY) and drives the 1RY relay. When the servo driver is in alarm, the signal will be invalid, and 1RY will de-energized, therefore the 1KM will be turned off and the main power supply will be cut off.

The function description of the host I/O signal connector X1 is shown in Table 3-6. The servo driver has three control modes, i.e. position/speed/torque control modes (selected by the driver parameters). In the position control mode, the CNC outputs the pulse signal PULS and the direction signal SIGN to control the rotation angle (旋转角 xuán zhuǎn jiǎo) and the rotation direction of the motor, and the closed-loop control is done by the driver. Under the speed/torque control mode, the speed and torque of the motor is

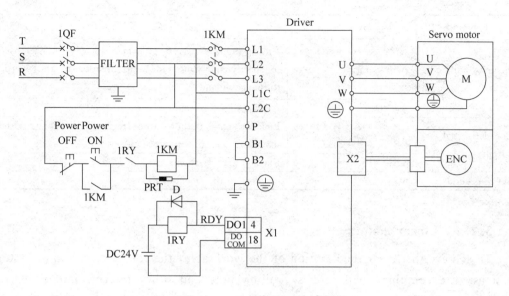

Figure 3-35   Main circuit of PMSM servo

specified by the analog signal, and the actual motor rotation angle is obtained by reading the encoder signals. In this mode, the CNC is responsible for the position(位置 wèi zhì) closed-loop control. It will compare the actual rotation angle and the command angle of the motor to adjust the speed/torque command output. In the tuning of CNC machine tools, the driver parameters of each feed-axis motor should be set, including the control mode, moment of inertia, motor gains, the command/feedback pulse frequency (electronic gear), friction force/ torque compensation(转矩补偿 zhuǎn jǔ bǔ cháng), sequential control actions and so on, in order to achieve a good matching of the driver, motor and feed mechanism, as well improve the control accuracy and response speed.

Table 3-6   Function description of host I/O signal terminals

Function	Symbol	Description
General-purpose digital input	DI1—DI5	Photoelectric-isolated programmable input signals, the functions are defined by the parameters
	COM+	Power supply of the general-purpose digital input (DC 12—24V)
General-purpose digital output	DO1—DO3	Photoelectric-isolated programmable output signals, the functions are defined by the parameters
	DOCOM	Common terminal for the general-purpose output
Position pulse command	PULS+, PULS−, SIGN+, SIGN−	High-speed photoelectric-isolated input signals, which can be set in different formats: (1) pulse train/direction; (2) CW/CCW pulse trains; (3) 90°-phase-difference 2-phase pulse

continuous

Function	Symbol	Description
Analog command input	AS+, AS−	Analog input signal, ranging in −10 V—+10 V, used to specify the magnitude and direction of the speed/torque command.
	AGND	Analog signal ground
Encoder signal output	OA+, OA−, OB+, OB−, OZ+, OZ−	Encoder signal output (after frequency division). Used for the feedback of motor angle.
	GND	Encoder signal ground
Shield ground	SG	Connect to the shield of the cable

3.2.4 Linear motors

To achieve the linear feed motion of the worktable, the rotary servo motor system must use the coupling, ball screws, rolling nuts and other intermediate parts for movement conversion and transmission. With the development of high-speed and high-precision machining technology, it requires the feed mechanism to be able of high speed, high acceleration, high accuracy and high rigidity. However, the traditional method using a rotary motor to drive the ball screw becomes difficult to meet the requirements of high-speed and high-precision machining(高精度加工). In order to improve the performance of linear motion system, the direct driving technology using the linear motor has been developed in recent years. Linear motor is a device that can convert the electrical energy into the mechanical energy of linear motion, which eliminates the intermediate transmissions from the motor to the worktable to achieve the "zero transmission", and significantly improves the CNC machine speed, precision and reliability.

1. Classification and characteristics of linear motors

According to the working principle, the linear motors can be classified into three categories: the linear stepping, the linear DC and the linear AC motors. The last ones are the most commonly used in the feed servo systems, including AC linear induction motors (asynchronous) and AC linear synchronous motors.

The AC linear synchronous motor can be classified further as the electromagnetic, permanent-magnet, variable-impedance, hybrid and superconductor types. The permanent-magnet synchronous linear motor needs a permanent-magnet steel to be laid on the machine tool, which makes the machine assembly, usage and maintenance(维护) difficult. The induction-type AC linear motor only generates magnetism when the power is on, thus avoiding the above disadvantages, so it is convenient for installation and maintenance, and the manufacturing cost is low. But the permanent-magnet synchronous linear motor is superior to the induction AC linear motor in the thrust force, efficiency, controllability(可控性) and stability. With the continuous improvement of rare-earth permanent-magnet materials, the permanent-magnet synchronous linear motor is becoming the mainstream.

In addition, according to the structure, the linear motors can be divided into the flat-type (including unilateral/bilateral types, etc.) and the cylindrical-type.

The machine tools employing linear motor(直线电动机)servo system removes the intermediate components in the transmission chain, which simplifies the machine structure and achieves direct driving to avoiding the problems such as the clearance, friction, insufficient rigidity(刚性) of ball screw and therefore improves the overall performance of the machine. The linear motor has the following advantages:

(1) Fast speed and short acceleration/deceleration time. The linear motor feed servo system can meet the requirement of high-speed machining, which can achieve a maximum feed speed of 60—100 m/min or higher, and can bear a maximum acceleration of 2—10 g.

(2) Fast response. Due to the elimination of the mechanical transmission parts with long response time, the whole response time of the feed system is shortened, and the dynamic performance is greatly improved.

(3) High precision. Due to the cancellation(消除) of the intermediate mechanical transmission structure, the transmission error is reduced and the precision of the feed system is improved. In addition, the linear motor needs linear displacement detection components to realize the closed-loop control, which can directly compensate the actual position error of the worktable and greatly improve the positioning accuracy of the machine tool.

(4) High transmission stiffness and smooth thrust force. Through the structure of "zero transmission", the stiffness reduction problem caused by the intermediate transmission mechanism(传动机构) is avoided. Meanwhile, by optimizing the layout of the machine tool, the force distribution of the feed system is nearly even and symmetrical, and the output thrust force is smooth.

(5) Low running noise. The linear motor eliminates the friction(摩擦) of the intermediate transmission components, and the rolling guide rail(滚动导轨)or the magnetic suspension guide rail can be adopted to further reduce the motion noise.

(6) Unlimited travel length. The length of the guide rail can be extended by append additional guide rail components to realize infinite extension(扩展).

(7) High efficiency. Due to the elimination of the intermediate transmission parts,

the friction loss in the transmission is reduced, and the mechanical efficiency is improved.

The application of linear motors in CNC machine tools should pay attentions to the following problems.

(1) Magnetic isolation protection. The magnetic field of the rotary motor is closed, while the magnetic field of the linear motor is open. The distance between the linear motor and the worktable is very close, which is prone to attract the magnetic materials such as the workpiece(工件 gōng jiàn), chips and tools and thus disturb the normal operation. So, it is necessary to carefully consider the magnetic isolation protection in the design.

(2) Heating problem. Because the linear motor is installed with in the machine tool with poor heat dissipation condition, the heat of the driving component will directly cause the thermal deformation of the guide rail of the machine tool. Therefore, the cooling problem should be solved in the design.

(3) Load disturbance. Since the linear motor directly drives the worktable using the fully closed-loop control, the change of the inertia of the workpiece load and the change of the cutting load are the disturbance to the control system. Therefore, in the design of the machine tool structure and control system, the robustness against the load disturbance should be considered.

(4) Balance of the vertical axis. For vertical installation of the linear motor, it is necessary to have a balance weight block(平衡重块 píng héng zhòng kuài) and a power-off self-locking device. Besides, the problem of the gravity should be tackled in the servo driver.

2. The structure and working principle of linear motors

From the structural point of view, the linear motor is equivalent to "cut" through the radial direction of a rotary motor and "unroll" it around its axis to form a plane structure. Figure 3-36 and Figure 3-37 are the structural evolution of the linear induction motor and linear PMSM respectively. For the linear motor, the part evolved from the original rotary motor stator is called the "primary", and the part evolved from the rotor is called the "secondary". In the drawing, the stator of the rotating motor is expanded into a flat primary part, and the rotor is expanded into a flat secondary, which results in a unilateral flat linear motor. If there are two primaries installed on the both sides of the secondary, it becomes a bilateral flat linear motor. In addition, the rotary motor can be evolved into a cylindrical structure to form a cylindrical linear motor.

In practical applications, the primary and secondary lengths are not equal, in order to ensure the constant coupling between the primary and the secondary in the full travel distance, so it can be divided into short-primary long-secondary structure and long-primary short-secondary structure. At present, due to the lower cost, the short primary structure has been more widely used, in which the fixed and moving parts are opposite to

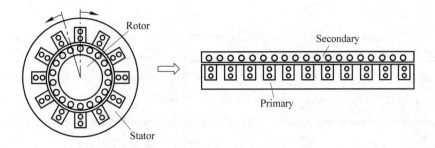

Figure 3-36 Induction-type asynchronous linear motor structure

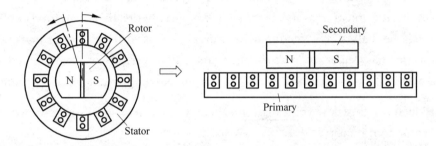

Figure 3-37 Permanent magnet synchronous linear motor structure

that in the rotary motors (i.e. the primary becomes the moving parts).

The working principle of linear motors is similar to that of the rotary motors. Take the short primary asynchronous linear motor as an example, when applying the multiphase-balanced AC in the primary windings, a sinusoidal magnetic field distributed along the linear direction will be established, as shown in Figure 3-38. The magnetic field is moving in a straight line according to the energizing sequence of the coil windings. The speed of the travelling wave magnetic field is the synchronized speed v_s. When the travelling-wave magnetic field moves, the conductive bars in the secondary will be cut by the magnetic field, the induced current will be generated, which produces the electromagnetic thrust with the travelling-wave magnetic field. Since the secondary is fixed, the primary will move in the opposite direction of the travelling-wave magnetic field by the electromagnetic force, and the speed is v. According to the principle of the induction motor, the moving speed is lower than the synchronized(同步的) linear speed v_s.

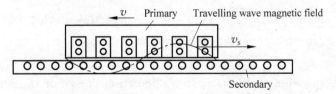

Figure 3-38 Working principle of the short primary induction-type linear motor

3. Application of linear motors

At present, the linear motor has been industrialized and applied in the high-speed, high-precision CNC machine tools. Since the first high-speed machining center using linear motors had been exhibited in the Germany Hannover in 1993, the speed, accuracy and reliability of linear motors have been continuously improved, and applications in the machine tools are becoming mature. At present, the application fields of linear motor include ultra high-speed die processing machine tools(模具加工机床), ultra-precision mirror processing machine tools, high-speed forming machine tools, etc. Most of the well-known CNC machine tool manufacturers have introduced high-end CNC machine tools driven by the linear motors. Although the cost of the linear motor is still higher than that of the ball screw feed mechanism and it is still in the primary stage of application, it has superior performance and high efficiency, which reduces the operating costs of the enterprise. With the continuous improvement of technology and the cost reduction, this new transmission scheme(传动方案) will become widely used in the machine tool industry.

3.3 Commonly used detecting elements

The detection device of the CNC provides the real-time position and velocity feedback information for the servo controller to achieve a closed-loop control system(闭环控制系统), which is an important part for the feed servo system. The smallest change that a detection device can measure is called the resolution. The resolution is not only related to the sensor itself, but also depends on the measurement circuit and the signal processing method.

The position detection devices include linear and rotary types. For the straight moving worktable, the linear measuring device can measure the displacement of the worktable directly, and it is not affected by the intermediate transmission parts of the machine tool. If the rotary type detection device is used, it will become indirect measurement, and the measurement accuracy(测量精度) is affected by the precision of the transmission chain(传动链) between the detecting device and the worktable. Therefore, the precision compensation method is often used to correct the measurement result.

For the CNC machine tool, the position detection device of the servo system needs to satisfy the characteristics of high precision, fast response, high reliability, low cost, easy maintenance, and so on. Nowadays, the commonly used position detection devices

on machine tools include rotary encoders, linear encoders, magnetic scales(磁栅尺),
_{cí shān chǐ}
resolvers, inductosyns, etc. The rotary encoders and linear encoders are mostly widely applied.

3.3.1 Rotary encoders

The rotary encoders belong to rotary position detecting devices, which convert the angular displacement to the pulse signal to indicate the change of the angle or the speed of the rotating shaft. According to the measuring principle(测量原理), rotary encoders can be divided into the contact type, the electromagnetic-induction type, photoelectric and other types. Among them, the photoelectric encoder is the noncontact type, which has no friction(摩擦), small driving force, fast response and high reliability, thus become the most popular product. According to the different coding schemes, the rotary encoders can be also divided into incremental and absolute types.

1. Incremental encoders

The rotary incremental encoder(增量编码器) outputs one pulse when the shaft rotates for a certain angular displacement, and the servo system of the CNC counts the number of pulses to obtain the corresponding angular displacement. The working principle of the rotary incremental photoelectric encoder is shown in Figure 3-39. The encoder consists of a light source, a lens, a graduated disk, photosensitive elements and a signal processing circuit. The graduated disk is usually made of the chromium-plated glass plate that is manufactured by photolithographic processes. The disk is engraved with circular graduations, which has equally distributed lines along the radial direction that segment the disk into a series of transparent and opaque zones. The number of lines is usually 500—5,000. The disk is connected with a rotating shaft, which could drive the disk in two directions. The light emitted by the light source is refracted into parallel light, then pass through the transparent lines on the disk, and finally received by the photosensitive elements. When the disk rotates, due to the effect of lines, the photosensitive element receives a light signal alternated in bright and dark, and generates an electronic signal which is similar to the sine waveform, as shown in Figure 3-40(a). Because a single output signal cannot distinguish the rotation direction, usually two photosensitive elements in different positions are employed to receive the light signals. The two signals have a 90° phase difference, known as the phase A signal and the phase B signal. In one direction, the A signal is ahead of the B signal by 90°, while in the opposite direction, the B signal is ahead of the A signal by 90°. Therefore, by comparing the relationship between A and B, we can obtain the direction of rotation(旋转). The output of the photosensitive elements(感光元件) are shaped and amplified by the signal processing circuit, and is converted into square-wave pulse trains, as shown in Figure 3-40(b). The relative angular displacement can be obtained by the bidirectional counting of A and B pulse signals. In order to improve the resolution, each rising or falling edge of A and B can be counted, so that a bright/dark cycle(周期) can generate 4 counting signals, which increases the resolution by 4 times. In addition, the speed of rotation can be evaluated by counting the number of pulses per unit time. Usually, a rotary encoder also generates a Z-phase pulse (also called index pulse or zero-point pulse) each revolution. The phase-Z signal can be combined with the origin switch of the machine tool to achieve accurate (准确的) return to the origin of the operation. And it also can be used for the absolute positioning signal of the spindle angle for realizing the thread cutting and spindle quasistop function.

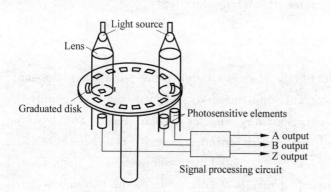

Figure 3-39 Working principle of rotary incremental photoelectric encoder

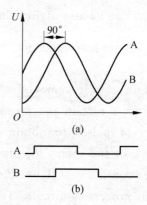

Figure 3-40 Incremental photoelectric encoder signal waveform

2. Absolute encoders

The incremental encoder cannot save the position information(位置信息) after power off, thus the feed servo system using the incremental encoder needs to return to the origin when it is powered on. In addition, the incremental encoder will produce position offset if it receives disturbing pulses. The absolute rotary encoder can generate the absolute angular position signal. Even after power off or disturbing pulses, the correct angular position can still be obtained. The working principle of the absolute encoder is similar to the incremental encoder except for the following two points: The absolute encoder is engraved with n-bit concentric code rings, and the combined n-bit angle code can be read from n photosensitive elements. The higher the encoder bits, the finer the resolution(解析度). Figure 3-41 is the disk of 4-bit absolute encoder, where the black and white areas corresponding to the opaque and transparent areas, respectively. Figure 3-41(a) is 4-bit binary disk, which has 4 code rings, and the rings from the inside to the outside correspond to 2^3, 2^2, 2^1, 2^0. The transparent area represents "0" and the opaque area represents "1". The 4-bit binary code divides the disk into 16 sectors (sector 0—15). For example, "1010" is the angular position "10" in the decimal number. The disadvantage of binary disk is that, more than one bit may "flip" in the adjacent angles (i.e. from 0 to 1, or from 0 to 1). Therefore, at the moment of crossing sectors, due to the flipping time error of each bit, the flip is not perfectly synchronized, which is prone to produce encoding error. An improved scheme(改进方案) is using the Gray code for encoding, as shown in Figure 3-41(b). In the Gray code, only one bit is flipped in the two adjacent sectors, thus avoids the synchronizing problem in the multibit flipping and improves reliability.

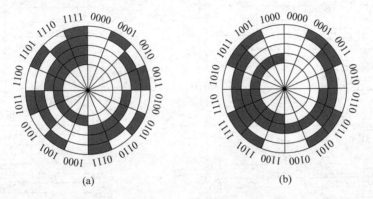

Figure 3-41 Absolute encoder with different formats
(a) Binary code; (b) Gray code

Compared with the incremental encoder, the position of the absolute encoder can be read directly from the encoder, and the actual position can be saved even after power off. However, the manufacturing process of absolute encoders(绝对编码器) is more complicated and the cost is higher. In recent years, a new type of absolute encoders has been developed, which no longer relies on the Gray code but uses multiple sine/cosine periodical coding tracks, and integrates the interpolation circuit to subdivide the signal, which can easily achieve a resolution of more than 17 bits or above. This new type of absolute encoders can improve the performance of the servo system.

3. Application of rotary encoders in CNCs

The following example is to illustrate the application of rotary encoder in the closed-loop control of feed motors. As shown in Figure 3-42, the feed servo system of the CNC employs the PMSM as the driving element, and the incremental rotary encoder is used as the feedback sensor, thus it makes a semiclosed-loop system. The incremental rotary encoder and the servo motor are coaxially installed, and the output signals of A and B are different in 90°. The servo driver counts and analyzes the pulse signals to obtain the actual angular displacement of the motor for feedback. The Z signal is generated by the encoder on every revolution. In the encoder of PMSM, the U, V, W phase signals are different by 120° in phase, which are used for the rough positioning of the rotor angle when the motor is power on in order to achieve a smooth start. The encoder is powered by a 5 V power supply. To improve the anti-interference ability, the encoder uses differential driving output, and uses twisted-pair cables to transmit the differential signals. In addition, the shielding layer of the signal cable is grounded by the dedicated FG terminal of the servo driver to further suppress the external interference(干扰﹝gān rǎo﹞).

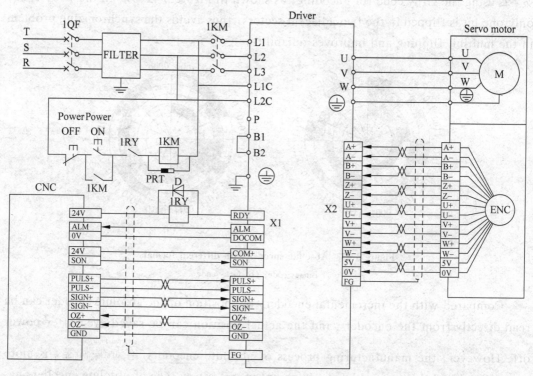

Figure 3-42 Semiclosed-loop CNC feed servo system based on incremental encoder

Now let us see the control principle of the above-mentioned servo feed system. The servo driver is operated in the position control mode, that is, the motor driver realizes the closed-loop control of position, velocity and current. The CNC sends the pulse/ direction signal (PULS/SIGN) to indicate the relative displacement and the direction of the feed. The driver monitors the position command from the CNC and the actual position feedback from the encoder in real-time, and compares the two to determine the velocity command to achieve the position closed-loop control. In addition, the Z signal is transmitted(传输 chuán shū) to the CNC through the OZ pin by the servo driver, which is used for the reference-point return function. All of the above signals use differential signals and twisted-pair cable for transmission. SON is the servo-enable(启用的 qǐ yòng de) signal, used to enable the output three-phase current to drive the servo motor and achieve closed-loop control. ALM is the servo alarm signal to the CNC, which becomes valid when the servo driver is in the alarm state. The SON and ALM are digital signals of 24 V level isolated by photocouplers. The signal cable between the CNC and the servo driver is shielded to prevent the external interference.

3.3.2 Linear encoders

The optical linear encoder is a detecting device for measuring the linear displacement. The displacement of the worktable can be directly measured(直接测量 zhí jiē cè liáng) in the CNC machine tool and as feedback given to the CNC by pulse signals. The linear encoders used in the machine tool utilize the transmission or reflection of light, and apply the theory of moire fringes. Nowadays, with the laser technology and the subdivision circuit, the resolution of the linear encoder can achieve up to 0.1 μm. In addition, the response speed of the linear encoder is very fast, and the measuring distance is very long, thus the linear encoders have been widely applied in the CNC machine tools. However, the high-precision linear encoders are fairly poor in terms of high manufacturing cost, complicated installation and being sensitive to vibrations and oil pollutions(污染 wū rǎn).

1. The structure of linear encoders

The optical linear encoder is composed of a light source, a lens, a scale grating, an indicating grating, a photosensitive element and a signal processing circuit. The scale grating is usually mounted on a moving part of the machine tool, such as the worktable. And the indicating grating is usually mounted on a fixed part of the machine tool, such as the base. When installing, it is necessary to ensure the parallelism between the scale grating and the indicating grating. The optical linear encoders can be classified into two types: the transmission type (Figure 3-43) and the reflection type (Figure 3-44). Take the transmission-type encoder as an example, the light emitted by the light source is refracted into parallel light, then vertically pass through the scale grating and indicating grating to form the moire fringes, and finally received by the photosensitive element. The brightness of the moire fringes received by the photosensitive element is approximately(近似的 jìn sì de) sinusoidal distributed. It usually employs two photosensitive elements to generate two signals with 90° phase difference. The signal of the photosensitive element is amplified, subdivided and shaped by the signal processing circuit to generate the pulse output signal for indicating the displacement and direction. Usually, the scale grating is provided with periodical zero reference marks, which can generate zero pulse signals. In the reflection-type linear encoder, the scale grating is often made of steel tape, and the parallel light is illustrated on the grating at a certain angle. The reflected light pass through the indicating grating(光栅) to form the moire fringes(莫尔条纹), then received by the photosensitive element. Generally, the length of reflection-type linear encoder is longer than that of transmission type, but the resolution is relatively low.

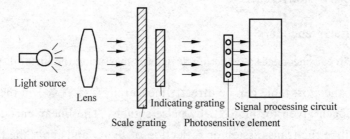

Figure 3-43   Principle of the transmission grating measurement system

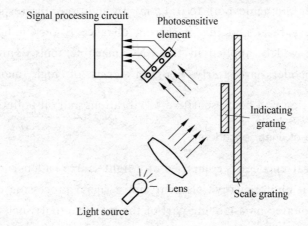

Figure 3-44   Principle of the reflective grating measurement system

2. Principle of the optical linear encoders

The moire fringe effect is utilized in the optical linear encoder to magnify the displacement. As shown in Figure 3-45, the scale grating and the indicating grating have a same pitch τ. Stack the two grating with a very close distance and offset the indicating grating with a small angle θ. Under the parallel light, the gratings form an equally spaced moire fringes alternate with bright and dark strips. Because θ is very small, the direction of the moire fringes is approximately perpendicular to the grating direction, so it is called the transverse moire fringes. The distance between two adjacent bright (or dark) zones in the moire fringes is called the width or pitch of the moire fringe W. According to the geometric relationship, the width of the moire fringes is

$$W = \frac{\tau}{\sin\theta} \approx \frac{\tau}{\theta} \tag{3.11}$$

Therefore, the moire fringes magnify the grating pitch τ by $\frac{1}{\theta}$ times, and the smaller the θ, the larger the magnification times. When the two gratings move relatively to a pitch of τ, the moire fringes will correspondingly move a distance of width W. Since the width of the moire fringes is much larger than the grating pitch, it is beneficial to improve the measurement resolution. By counting the number of the moved strips on moire fringes, the number of travelled grating pitches can be obtained. The relative moving direction of the gratings can also be reflected in the moving direction of the moire fringes. In addition, the moire fringes are produced by the effect of multiple gratings, which has the average effect to average the pitch errors of adjacent lines, thus improves the measurement accuracy.

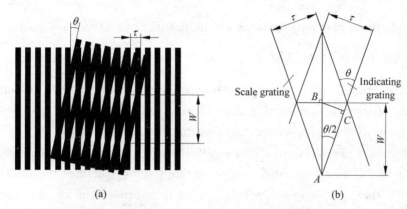

Figure 3-45 Moire fringe effect principle

(a) Forming of moire fringes; (b) Parameters of moire fringes

Chapter 4　Mechanical structure of the CNC machine tool

　　The mechanical structure(机械结构) of a CNC machine tool is similar to that of a general machine tool and it consists of the following parts: main drive system(主传动系统) and spindle components (making the tool/workpiece have main cutting motion); feed drive system(进给传动系统) (making the workpiece/tool have feed motion and achieving positioning); support components, such as the bed(床身), column(立柱), slide seat and worktable (supporting the machine tool); and other auxiliary devices, such as hydraulic, pneumatic, lubrication, cutting fluids.

　　Although there are many structural similarities between the CNC machine tool and the general machine tool, the modern CNC machine tool is neither the traditional one equipped with the CNC system, nor is via local improvement on the basis of the traditional. Because a general machine tool has the disadvantages of low rigidity, poor vibration resistance and large thermal deformation, it cannot meet the machining requirements for high speed, high precision, etc. The modern CNC machine tool, especially the machining center, has undergone great changes in the structure of components such as large basic components, main drive system, the feed system, the tool system and auxiliary device, as well as the overall layout, external modeling, etc., forming the unique mechanical structure of the CNC machine tool.

4.1　Requirements for the CNC machine tool for mechanical structure

　　The CNC machine tool requires high static/dynamic stiffness(静动刚度) and small thermal deformation(热变形) of the supporting components, small friction(摩擦) between the moving components and the humanized(人性化的) operation.

　　1. Supporting components with high static/dynamic stiffness

　　1) Concept of static/stiffness

　　Machine tool stiffness(机床刚度) refers to the ability of the structure of a machine tool to resist deformation(抵抗变形). According to the nature of the load, the stiffness of the machine tool represented under the action of static force (gravity, interaction between mechanisms) is called "static stiffness". The stiffness represented under the action of dynamic force (cutting force) is called "dynamic stiffness".

A remarkable feature of a CNC machine tool is that it can be processed at high speed and high precision(高速高精). The methods of increasing cutting speed, adopting automatic tool changing system and automatically changing pallets can be used to achieve this goal. These methods have greatly increased productivity as well as the load and running time of the machine tool. If the stiffness is insufficient, under the action of gravity and cutting force, the mechanical deformation of the components and members of the machine tool will cause the deviation of the relative displacement between the cutting tool and the workpiece, thus affecting the machining precision and surface quality. Therefore, a CNC machine tool has higher dynamic/static stiffness than a traditional machine tool.

2) Measures to increase dynamic/static stiffness

Main factors which affect machine tool stiffness are the stiffness of the members and the components themselves(部件本身) and the contact stiffness in-between.

In order to improve the stiffness of the large components of a machine tool, a bed with closed interface is used, as shown in Figure 4-1, and a weight hammer or hydraulic balance(重锤或液力平衡) (as shown in Figure 4-2) is used to reduce the machine tool deformation caused by the change of the position of moving components.

Figure 4-1　Closed integrated box

In order to improve the stiffness of the spindle of the CNC machine tool, the three-support structure is often adopted and the double-row short cylindrical roller bearing and the angular contact radial thrust bearing are used to reduce the radial and axial deformation of the spindle. In the CNC machine tool, the spindle-end-overhang cantilever

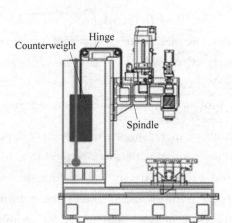 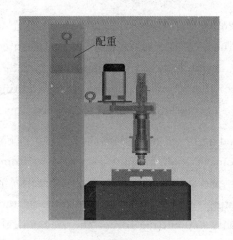

Figure 4-2 Spindle weight balance structure

length should be as short as possible to reduce the bending moment.

In the machine tool, in order to improve the contact stiffness(接触刚度 jiē chù gāng dù) of various parts and increase the bearing capacity, scraping and grinding methods are used to increase the contact points per unit area and apply a large enough preload between the joining surfaces to increase the contact area. These measures can effectively improve the contact stiffness.

The measures commonly used to improve the dynamic stiffness(动刚度 dòng gāng dù) include improving the stiffness of the system, increasing damping(阻尼 zǔ ní) and adjusting the natural vibration frequency of components. Experiments show that increasing the damping coefficient is an effective way to improve vibration resistance. The welded structure of steel plates can not only increase the static stiffness and reduce the weight of the structure, but also increase the damping of the member itself. Therefore, in recent years, the body, column, beam and worktable with the welded steel plate structure have been adopted for CNC machine tools. Sand-sealed castings are also favorable to vibration attenuation and have good effect on improving vibration resistance(抗振性 kàng zhèn xìng).

In large cavities filled with damping material such as mud core and concrete(泥芯和 ní xīn hé 混凝土 hùn níng tǔ), the vibration energy is dissipated during vibration due to high friction. The damping layer method can also be used, that is, a viscoelastic material with high damping and elasticity is sprayed on the surface of large parts, the thicker the coating(覆盖层 fù gài céng), the greater the damping.

In recent years, there have also been ways using DCG technology and box-in-box construction to increase stiffness.

The drive at the center of gravity(DCG 重力驱动 zhòng lì qū dòng) technology is based on the theory of mechanical motion dynamics. DCG as shown in Figure 4-3, uses two ball screws to fix the center of gravity of the worktable and uses twin drives to suppress vibration and improve dynamic stiffness. (Vibration is the biggest affecting factor of high speed and high precision.) It is the best way to connect the centers the two ball screws and the middle point of the line between the two centers coincides with the center of gravity of the moving object. Similar to pushing an object, pushing in the middle will cause the object to move smoothly without turning.

Figure 4-4 shows a box-in-box structure. Its spindle box installed in the motion box can move smoothly with good stiffness. In addition, the table-in-table structure that the C-axis worktable is configured in the B-axis worktable realizes stable processing precision with the high-stiffness structure.

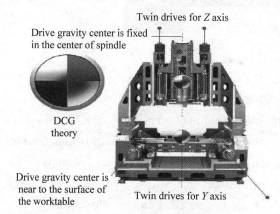

Figure 4-3 DCG principle

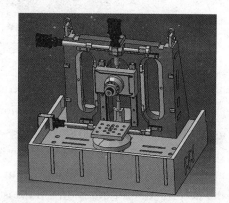

Figure 4-4 Box-in-box structure

2. Measures for supporting components to reduce thermal deformation

Under the impact of external/internal heat sources such as cutting, friction and so on, the components of the machine tool will undergo different degrees of <u>thermal deformation</u>(热变形), as shown in Figure 4-5. (When the feedrate is 10m/min, roller screw becomes heat and temperature will be from 25℃ to 40℃ during multiple reciprocating movements. Without compensation, there is a serious problem with the workpiece.) Thermal deformation will destroy the relative kinematic relationship between the workpiece and the tool, and it is difficult for operator to correct, thus may seriously affect the workpiece processing precision. In order to reduce thermal deformation, the following measures are usually taken in the structure of CNC machine tools.

1) Reduce heat

The internal heat is the main heat source to produce thermal deformation and the heat source should be separated from the main unit to the greatest extent.

2) Control temperature rise

After taking a series of measures to reduce the heat source, the thermal deformation will be improved. However, it is usually very difficult or even impossible to completely eliminate the internal/external heat sources of the machine tool. Therefore, good heat dissipation and cooling system must be used to control the temperature rise to reduce the heat source. One of the more effective methods is to forcibly cool the hot parts of the machine, such as cooling of cutting parts with "Multi-Nozzle, large flow, forced cooling procedures".

Figure 4-5　The temperature-rising status for roller screw

3) Improve the machine tool structure(改善机床结构)

　　Under the same heating conditions, the machine tool body also has a great impact on the thermal deformation. For example, the CNC machine tool adopts the double-column mechanism instead of the single-column one. Due to the left-right symmetry, the main axis heated by the double-column mechanism generates a vertical translation and the deformation in other directions is small, while the vertical axis of movement can be easily compensated with a coordinate correction amount. For the spindle box of the CNC machine tool, we should try to make the spindle thermal deformation occur in the vertical direction, that is, in a non-error-sensitive direction, the cutter cuts in. This minimizes the effect of spindle thermal deformation on the machined diameter. In respect of the structure, we should also reduce the length of the spindle as much as possible to reduce the total amount of thermal deformation, while the temperature should be consistent before and after the headstock to avoid tilting after deformation of the spindle.

4) Pre-stretch the lead screw(预紧丝杠)

　　The ball lead screw of a CNC machine tool is often subject to large expected load, high rotational speed and poor heat radiation, so it can easily get heated. Its consequence is serious, especially in an open-loop system, it can degrade the feed system's positioning precision. At present, some machine tools use the pre-pull method to reduce the thermal deformation of the ball lead screw.

5) Compensate thermal displacement(补偿热位移)

　　The thermal deformation that cannot be eliminated after the measured mentioned above can be corrected by the compensation pulse transmitted from the CNC system according to the measurement result.

3. The friction between the moving parts of a CNC machine should be small and the clearance of the drive system should be eliminated

The displacement of the CNC machine worktable (or carriage) is based on the pulse equivalent as a minimum unit and is usually required to move (such as alignment and tool calibration) at very low speed. In order to enable the worktable to accurately respond to the instructions of the CNC device, corresponding measures must be taken to improve the friction characteristics between the moving parts. In the feed system of a CNC machine tool, the static and dynamic friction ball screws replace the sliding screw, and the hydrostatic guide rail, rolling guide rail or plastic guide is used. As a result, high-speed feed will have no vibration, low-speed feed will have no crawling, the sensitivity is high; and the machine tool can work continuously for a long time under heavy load and achieve high wear resistance and good precision retention.

The machining precision of a CNC machine tool (especially the CNC machine tool in open-loop system) depends to a great extent on the precision of the feed chain. In addition to reducing the machining errors between the drive gear and the ball screw, another important measure is to use gapless auxiliary transmission.

4. The structure of the CNC machine tool should be humane(人性化的)

In the single workpiece processing by a CNC machine tool, auxiliary time (non-cutting time) occupies a larger proportion. To further improve its productivity of a machine tool, we must take measures to minimize auxiliary time. There have already been many CNC machine tools with multiple spindles, multiple turrets and the automatic tool changer to reduce tool change time. At the same time, this also saves manpower and reduces the labor intensity of people. In addition, the color of the machine, the position of the machine operating panel and the height must be humane.

4.2　Current common types and layout of CNC machine tools

4.2.1　CNC Lathes

In the view of the structure, CNC lathe beds mainly are the horizontal(卧式), the slant(倾斜式), the horizontal with taper slide(卧式带斜滑板) and the vertical(立式).

The horizontal bed has good workmanship and the guide rail surface is easy to process, as shown in Figure 4-6. It is equipped with a horizontal turret which can improve the movement precision of the turret. Generally, it can be used in the layout of large CNC lathe or small-precision CNC lathe and it is mostly used for the reconstruction of common CNC lathes. However, the horizontal bed is difficult to discharge chips because of the small space in the lower part. In terms of structure dimensions, the horizontal placement of the turret makes the lateral dimension of the slide plate longer, thus increasing the structural dimension of the machine tool in the width direction and affecting the appearance of the machine tool.

Figure 4-6　Horizontal bed CNC lathe

The horizontal bed with taper slide means that its body is equipped with a taper slide and a slant guide rail shield. Firstly, this type of layout keeps the characteristics of good workmanship in the horizontal bed. Secondly, the width of the machine tool is smaller than the horizontal configuration of the slide; in addition, this layout facilitates chip discharge.

The slant bed (shown in Figure 4-7) can be divided into the 30°-/45°-/60°-/75°-/90°-slant (known as the vertical) according to the inclination of the guide rail relative to the ground. When the inclination angle is small, it will be inconvenient to discharge chips; when the inclination angle is large, the guide rail will have a bad guide, the force condition will be also bad. The inclination angle will also directly affect the ratio of the height of the machine tool to its width in relation to external dimensions. Considering all of the above factors, the 30°-/45°-slant beds are mostly used for small CNC lathes, the 60°-slant for medium CNC lathes and the 90°-slant for large CNC lathes.

In general, small and medium CNC lathes mainly adopt slant bed or flat bed taper slide structure. The advantages of these two layouts are as follows:

(1) Easy to discharge chips and the coolant and able to reduce bed body thermal deformation.

When a workpiece is processed, the chips and the cutting fluid can fall from the front of the bevel (i.e. one side of the bed) to the chip ejector in front of the bed and then the chips are discharged by the chip ejector. This layout has good chip removal and

Figure 4-7 Slant bed

heat radiation, thus reducing the heat generated by cutting in the bed, reducing the thermal deformation of the bed and finally ensuring the machine tool to keep the machining precision better.

(2) No need to open chip discharge holes in the bed body and able to improve bed body stiffness.

Because there is no need to open the chip discharge holes in the bed body, the whole structure of closed section can be adopted to improve the stiffness of the bed.

(3) Good look and convenient operation.

In case of slant bed or guide rail, the hot iron chips will not accumulate on the guide rail and it is easy to install automatic chip ejector. The bed body is easy to operate, easy to install the loading/unloading manipulator, so as to realize the automation of the single machine. In addition, the lathe has a concise and good shape; it covers a small area and can realize sealed protection easily.

4.2.2 Machining centers

Since 1959, there have been various types of structures for machining centers depending upon the form of the bed and worktable, the direction of the axis of the spindle axis and the movement of the coordinate axis and the linked coordinate axis.

Based on the bed and worktable, machining centers can be divided into the one with T-shaped bed layout (see Figure 4-8) and the one with cross-shaped worktable layout (see Figure 4-9). Depending on the direction of the axis of the spindle axis, machining centers can be divided into the horizontal and the vertical ones. While on the bases of movement coordinate axis and the linked coordinate axis, machining centers can be divided into the 3-axis 2-linkage, the 3-axis 3-linkage, the 4-axis 3-linkage, the 5-axis 4-linkage and the 6-axis 5-linkage, etc.

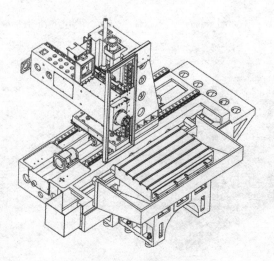

Figure 4-8　T-bed layout

Figure 4-9　Cross-shaped layout

1. Horizontal machining center（卧式加工中心）

Its spindle axis is parallel to the worktable. It usually adopts the mobile column (the supporting part of the spindle box bearing not only the force in all directions but also the relevant torques), the worktable does not rise and fall, and the T-shaped bed is adopted (see Figure 4-8). The machining center adopts T-shaped bed layout with the worktable supported on the bed, and it also has good stiffness and strong bearing the capacity of the worktable. The casting and processing properties of the split T-shaped bed are greatly improved, but the connecting position should be fixed by the positioning key and the special positioning pin, and the rigidity and precision should be ensured by securing with the large bolt. The stiffness and precision of the integrated T-shaped bed are better, but its casting and processing performance are worse.

The vertical columns of the horizontal machining centers generally adopt the structure of double columns, and the spindle box moves up/down along the guide rail between the two columns. This structure has good thermal symmetry and high stability.

The horizontal machining center generally has 3-axis linkage. Usually, it adopts the linkage of three linear coordinates X, Y and Z and a rotary coordinate B graduation. It can finish the machining of four sides under one clamping, and is especially suitable for machining the parts of box type.

2. Vertical machining center（立式加工中心）.

Compared with the horizontal machining center, the vertical machining center has the advantages of simple structure, small floor space and low price. The medium and small vertical machining centers generally use a cross-shaped worktable layout (see

Figure 4-9) and the fixed column structure, as the spindle box is suspended on one side of the column and usually uses a square cross-section frame structure, torx or grillage stiffeners, to enhance torsional stiffness, and the column is hollow to place the balance of the spindle box counterweight. The spindle box (mobile Z-axis) is equipped with a balance hammer or a balance pressure (hydraulic) cylinder designed for central guide so that the counterweight will not cause sloshing even when moving at high speed. Accurate proportion of counterweight to spindle weight can obtain the best processing characteristics. And the Z-axis drive motor has good load characteristics.

The vertical machining center usually also has three linear motion coordinates, its slide plate and worktable are used to realize the movement of two coordinate axes, i.e. X-axis and Y-axis, in the plane and the movement of the spindle in the Z direction.

lóng mén jiā gōng zhōng xīn
3. Gantry machining centers(龙门加工中心)

The gantry machining center, as shown in Figure 4-10, refers to the one in which the spindle axis is perpendicular to the worktable. It is mainly suitable for machining large parts.

Figure 4-10 Gantry machining center

Depending on whether the gantry can move, the gantry machining center can be divided into the mixed gantry and moveable worktable machining center (accounting for above 90%) and the moveable gantry machining center (also called bridge-type machining center). Bridge gantry mill is characterized by small floor space, large bearing capacity and up to 20 m travel of gantry frame, and it is convenient for the machining of specially long or heavy workpieces. Depending on whether the crossbeam moves on the column, the gantry machining center can be also divided into the moving-beam gantry machining center and the fixed-beam types. The gantry machining center with the crossbeam on the elevated bed is called the elevated-type one.

The machining of large and complex parts usually requires a lot of accessory heads. Accessory heads are specially designed according to the processing requirements of the workpiece, and are generally divided into right/extended/special-angle/universal heads, etc.

4. Five-axis machining center（五轴加工中心）

The five-axis machining center is generally composed of three linear axes (X-, Y-, Z-axes) and two rotary axes, enabling five-axis linkage, that is, during the cutting process, the linear axis and the rotary axis can simultaneously cut according to machining requirements. Based on different combinations of rotary axes, the five-axis machining centers can be divided into the double-turntable (Figure 4-11 (a)), the double-swing-axis (Figure 4-11 (b)) and the turntable-swing-axis ones (Figure 4-11 (c)).

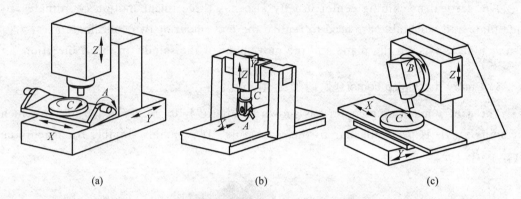

Figure 4-11 Five-axis machine tool structure
(a) Double turntable structure; (b) Double swing axis structure; (c) Turntable swing axis structure

4.3 Main drive system and spindle components

The role of the main drive system is to transmit the torque or power of the motor to the spindle components so as to realize the main cutting movement(主切削运动) of the workpieces or tools installed in the spindle. The main drive system of the CNC machine tool includes a spindle motor, a drive system and the spindle components. Compared with that of the ordinary machine tool, the structure of the main drive system of this machine tool is relatively simple because all or most of the variable speed function is implemented by the stepless speed regulation of the spindle motor, eliminating the need for complicated gear-shift mechanism; some main drive motors only have the two-or three-level gear-shift system to expand the range of motor stepless speed regulation. At present, the main drive motor of CNC machine tools almost does not use the ordinary AC asynchronous motor and traditional DC speed regulating motor, which are gradually replaced by the emerging AC variable-frequency servo motor and the DC servo speed-regulating motor.

1. Requirements of CNC machine tools on main drive system

1) <u>Wide speed-regulation range</u>（调速范围宽）

The speed-regulation range includes constant-torque constant-power speed-regulation ranges. At present, the speed-regulation range of the main axes of CNC machine tools is generally 100—30,000 r/min, and allows stepless speed regulation. The constant-power speed-regulation range is required to be as large as possible in order to use its full power at the lowest possible speed. Different machine tools have different requirements for speed-regulation range. The multi purpose universal machine tools require large speed-regulation range of the spindles, which should not only have the low-speed and large-torque function, but also have the high-speed, such as turning center. The special CNC machine tools do not need a larger speed-regulation range, such as for CNC gear machine tools, CNC drilling and boring machine ones in the mass production of automotive industry. Some CNC machine tools need to be able to process not only ferrous metal materials but also aluminum alloys and other nonferrous ones, which requires a wide speed-regulation range and ultra-high-speed cutting.

2) <u>Enough torque and power</u>（足够的扭矩和功率）

With enough torque and power, CNC machining can realize large torque at low speed and constant power at high speed so as to ensure high machining efficiency. Due to the increasing demand for high efficiency and the advancement of cutting tool materials and technologies, most CNC machine tools require sufficiently high power to meet the requirements of high-speed and high-strength cutting. The driving power of the spindles of the general CNC machine tools is 3.7—250 kW.

3) Small thermal deformation

The motor, spindle and drive parts all are heat sources. Low temperature rise and small thermal deformation are important indicators for the main drive system.

4) High rotation and motion precision of the spindle

The precision of the spindle rotation is the measurement of radial circular runout and end circular runout at 300 mm from the front end of the spindle under the conditions of no load and low-speed rotation after assembly. The two accuracies measured above when the spindle rotates are called motion precision. CNC machine tools require high precision of rotation and movement.

5) Good static stiffness and vibration resistance of the spindle

Due to the high machining precision and the high rotational speed of the spindle, the static stiffness and vibration resistance are required to be high. The dimensions of spindle journal, the type and configuration of the bearing, the preload value of the bearing, the mass distribution of the spindle assembly, the damping of the spindle assembly, etc., all affect the static stiffness and vibration resistance of the spindle assembly.

6) High wear resistance（耐磨性高）of the spindle assembly

The spindle assembly must have sufficient wear resistance so as to maintain long-term precision. All mechanical friction components, such as bearings, cone holes, etc., should have sufficiently high hardness and the bearings should also be well lubricated.

2. Driving mode of spindle

In actual production of a CNC machine tool, constant power is not always needed in the whole speed range. In general, the constant-power drive is required at high and medium-speed whereas the constant torque drive at low speed. In order to ensure that the spindle of CNC machine tools has greater torque and its speed range is as large as possible at low speed, some CNC machine tools are configured with gear shift based on AC or DC motor stepless speed changing（无级变速）to make it piecewise change accordingly, as shown in Figure 4-12(a),(b).

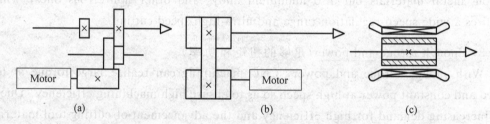

Figure 4-12 Three configurations of the main drive of CNC machine tool
(a) Gear shift; (b) Belt drive; (c) Built-in motor spindle drive structure

1) Main drive with speed gear (see Figure 4-12 (a))

This is the configuration mode often used by large and medium CNC machine tools and expand its speed range through a few pairs of gear transmissions. Since the constant power speed range of the motor above the rated speed is 2—5, when necessary to expand the speed range, the method of variable speed gear is used to do so. The displacement of slip gear is mostly realized with the hydraulic shift fork (as shown in Figure 4-13) or by driving the gear with the hydraulic cylinder.

2) Main drive with belt drive (See Figure 4-12 (b))

This drive is mainly used on the machine tools with high speed and small speed range. The motor itself can be adjusted to meet the requirements, without gear change, thus avoiding the vibration and noise from the gear drive. It is suitable for the spindle of high speed and low torque. The synchronous toothed belt (referred to as the timing belt) is commonly used.

3) Direct drive by the spindle motor

The spindle and the rotor of the motor are combined to make the structure of the spindle components more compact, light-weighted and with small inertia, effectively improving the stiffness of the spindle components and improving the response characteristics of the spindle to start/stop. At present, the spindle of the high-speed machine tool mostly adopts this way. This type of spindle is also known as the electrical spindle (电主轴). See Figure 4-12 (c).

Its disadvantage is that the heat generated by motor operation easily causes thermal deformation of the spindle. Therefore, temperature control and cooling are key problems in the use of internal motor spindle. As shown in Figure 4-14, the spindle assembly of a vertical machining center has a maximum motor spindle speed of 24,000 r/min.

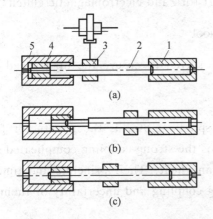

Figure 4-13　Work principle of three-position hydraulic shift fork

1,5—Hydraulic cylinder; 2—Piston rod;
3—Shift fork; 4—Sleeve

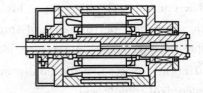

Figure 4-14　Built-in motor

3. Spindle speed regulation method

The spindle speed regulation (主轴调速) of a CNC machine tool is carried out automatically according to the control instructions. In order to meet the requirements of speed regulation and torque output of the main drive, the electromechanical method is commonly used in CNC machine tools, that is, the motor and the mechanical gear are used to change the speed at the same time. In this mechanism, the gear decelerating is for increasing output torque and the gear shift is used to extend the speed-regulation range.

1) Motor speed regulation (电动机调速)

The speed-regulating motor used in the spindle drive mainly includes DC motor and AC motor.

2) Mechanical gear shifting (机械齿轮调速)

CNC machine tools often adopt the combination of 1-4 gear shifts and stepless speed regulation, that is, the so-called piecewise stepless speed change. The mechanical gear deceleration is used to increase the output torque and the gear shift is used to extend the speed regulation range.

When a CNC machine tool is working, the spindle runs automatically according to the speed specified by the spindle speed instruction in the machining program. There are two kinds of command signals to control the spindle speed. One kind is analog or digital signal (code S in the program) to control the speed regulating circuit of the spindle motor; The other kind is switching signal (codes M41-M44 in the program) to control the executing mechanism of the mechanical gear automatic shift. The automatic gearshift actuator is an electric-mechanical conversion device and the commonly used automatic gearshift actuators are hydraulic shift fork and electromagnetic clutch.

4. High-speed spindle of the CNC machine tool

There are three technical problems to be solved to ensure the high speed of the spindle:

1) High-speed motor control technology

The control technology of a high-speed spindle motor is the key factor for the dynamic and static performance of spindle. It is the strong-coupling complicated system combing the mechanism, electricity, magnetic and heat. Their principle and time scale are different, so to solve the nonlinear, strong coupling and uncertainty of them is the key to control spindle well.

2) Development of high-speed bearing

The scheme of selecting ceramic bearing at high speed has been adopted in the machine tool of the machining center. The rolling body of the bearing is made of ceramic while the inner and outer rings are still made of bearing steel. The ceramic material is Si-N, which has the advantages of light weight (equal to 40% of that of the bearing steel), low thermal expansion rate (equal to that of 25% of the bearing steel), and high elastic modulus (equal to 1.5 times that of the bearing steel). The centrifugal force and inertia slip can be greatly reduced by using ceramic rolling body, and the spindle speed can be increased. The current problems are that ceramics are expensive, and the test data of life span and reliability are not enough and still need to be further tested and improved.

3) Research of cooling lubrication technology

In the past, the spindle bearings of the machine tools of machining centers mostly used grease lubrication. In order to meet the need of increasing the spindle speed, new cooling lubrication methods have been developed one after another.

Methods have been developed including oil and gas lubrication, injecting one and penetration roller one.

5. Spindle components(主轴部件)

Spindle components are the key components of machine tools and include the spindle support and drive parts installed on the spindle. The quality of spindle components directly affects machining quality. The spindle components of any machine tool should meet the following requirements: the rotation precision of the spindle; the structure, stiffness and vibration resistance of the components; the operating temperature and thermal stability; and the wear resistance and precision retention of the components. For CNC machine tools, especially for automatic tool changing ones, in order to realize the automatic loading/unloading and clamping of cutting tools on the spindle, there must also be the automatic clamping device for the cutting tools, the quasistop device of the spindle, the cleaning device of the spindle hole, etc.

1) Spindle end structure and shape(主轴端部的结构和形状)

The end of the spindle is used for mounting a tool/clamp for clamping a workpiece, and in terms of design, it should ensure accurate positioning, reliable installation, firm connection, and convenient loading/unloading, and be able to transmit sufficient torque. The structure and shape of the ends have been standardized, Figure 4-15 shows the several structural forms for common machine tools and CNC machine tools.

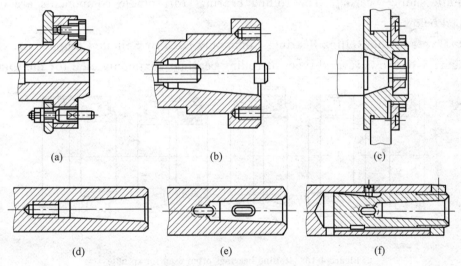

Figure 4-15　Structure of the spindle end

Figure 4-15 (a) is the spindle end of the lathe and the chuck is positioned by the short conical surface and flange face of the front end, and the torque is transmitted by the pull pin. The chunk is provided with fixing bolts. When the chunk is to be installed on the spindle end, thread such fixing bolts through the holes in the flange, turn the quick-release chunk plate to fasten the bolts and tighten the nuts to fix the chunk onto the spindle end firmly. The spindle is hollow and its front end has a Morse taper hole for mounting the core clamper or mandrel.

Figure 4-15 (b) shows the spindle end of the milling or boring machine tool. The milling cutter or arbor are located in the taper holes at 7:24 at the front end and tightened with tension rod from the rear end of the spindle and the torque is transmitted with the end surface key at the front end;

Figure 4-15(c) shows the end of the cylindrical grinding wheel spindle;

Figure 4-15(d) shows the structure of the inner circular grinder wheel spindle end;

Figure 4-15(e) shows the end of the drilling machine and ordinary boring bar. The cutter bar or tool is positioned by the Morse taper hole, the torque is transmitted through the first flat hole at the rear end of the taper hole, and the second flat hole is used to remove the tool. However, the CNC boring machine should use the form of (b). The 7:24 cone hole shown in the figure does not have a self-locking function, thus facilitating the pull out of the tool during tool replacement;

Figure 4-15(f) shows the structure of the CNC boring machine spindle end.

2) Support of Spindle Components

The spindle of the machine tool rotates in the support with the tool or fixture, and it should be able to transfer cutting torque and bear cutting resistance and ensure the necessary rotation precision. The spindle of a machine tool is usually supported by rolling bearing, and the spindle requiring high precision is supported by hydrodynamic or hydrostatic sliding bearing. The rolling bearings for spindle components are briefly described below.

(1) Types of the Rolling Bearings Commonly Used by Spindle Components.

Figure 4-16 shows several types of rolling bearings commonly used for the spindle.

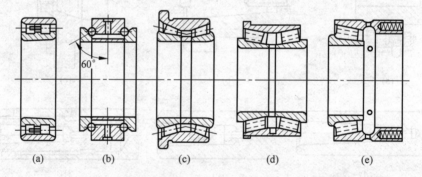

Figure 4-16　Rolling bearings often used for spindle

Figure 4-16(a) is a taper hole double-row cylindrical roller bearing with 1:12 taper holes in the inner ring. When the ring moves axially along the journal of the cone, the ring will expand to adjust the gap between the raceway. There are a large number of rollers and two rows of them are staggered; therefore, the bearing is characterized by large bearing capacity, good stiffness and high rotation speed. The inner and outer rings of the bearing are very thin; therefore, it is required that the spindle journal and the box bore have higher manufacturing precision, so as to avoid the distortion of the bearing raceway caused by the shape error of the journal and the box hole, which will affect the rotation precision of the spindle. The bearing can only withstand radial load.

Figure 4-16(b) shows a double-row thrust angular contact ball bearing with 60°

contact angle. The bearing is characterized by small ball diameter and large number of balls and can bear two-way axial load. The spacer sleeve in the middle can be ground thin to adjust the clearance or pre-tighten; the axial stiffness is high, and high rotation speed is allowed. With the double-row cylindrical roller bearing, the bearing is generally used as the front support of the spindle, and the outer diameter of the outer ring is negatively deviated to ensure that the bearing only bears axial load.

Figure 4-16(c) is a double-row tapered roller bearing, which has a common outer ring and two inner rings. Axial positioning is carried out on the box by the shoulder of the outer ring, and the box holes can be bored into through holes. The spacer sleeve in the middle can be ground thin to adjust the clearance or pre-tighten; the difference of one roller between the two rows of rollers can cause the vibration frequency to be inconsistent and thus improve the bearing dynamics obviously. This bearing can withstand both radial and axial loads and is usually used as the front support for the spindle.

Figure 4-16(d) shows the double-row cylindrical roller bearing with a shoulder, which is similar in structure to Figure 4-16(c) and can be used as the front support for the spindle. The roller is hollow, the cage is an integral structure. Filling the gap between the rollers, the lubricant flows from the hollow roller end face to the flange friction place, realizing effective lubrication and cooling. The hollow roller can produce small deformation when subject to impact load, and it can increase the contact area and has the functions of absorbing vibration and buffering.

Figure 4-16(e) shows a tapered roller bearing with a preloaded spring. The number of springs is between 16 and 20. The uniform increase and decrease of the springs can change the preload.

(2) Spindle bearing configuration(主轴轴承配置).

In practical application, there are the following three common configurations of the spindle bearing of the CNC machine tool, as shown in Figure 4-17.

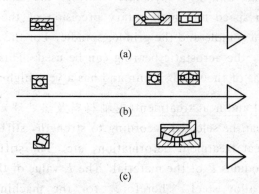

Figure 4-17　CNC machine tool spindle bearing configurations

As shown in Figure 4-17(a), the front support uses the double-row short cylindrical roller bearing and 60° angular contact double-row concentric thrust ball bearing, and the rear support uses the concentric thrust ball bearing pair. This configuration can make the spindle obtain greater radial and axial stiffness, and can meet the requirement of strong cutting of machine tools. It is widely used in the spindle of various kinds of CNC machine tools, such as CNC lathes, CNC milling machines, machining centers and so on. The rear support for this configuration can also be used for cylindrical roller bearings to further improve the radial stiffness of the rear support.

For the configuration as shown in Figure 4-17 (b), the front support uses multiple high-precision centripetal thrust ball bearings, and the rear support uses ball bearings. Although the stiffness is not as large as the spindle stiffness shown in Figure 4-17(a), this configuration improves the rotation speed of the spindle, and is suitable for the CNC machine tool requiring working at high rotation speed. At present, this configuration is widely used in vertical and horizontal machine tools for machining centers and it can meet the requirements of the wide range of rotating speed and high maximum rotational speed of this kind of machine tools. To improve the spindle stiffness of this configuration, four or more bearings can be used for the front support and two bearings for the rear.

As shown in Figure 4-17(c), double-row tapered roller bearings are used for the front support and single-row tapered roller bearings for the rear support. This configuration allows the spindle to withstand heavy loads (especially strong dynamic loads) and has high radial and axial stiffness, as well as good installation/adjustment. However, this configuration limits the maximum rotational speed and precision of the spindle and is suitable for the spindle of the CNC machine tool with medium precision, low speed and heavy load.

In order to improve the stiffness of the spindle assembly, the three-support spindle assembly is often used in CNC machine tools. Especially for the CNC machine tools with large span between front and rear bearings, the bending deformation of the spindle can be effectively reduced by auxiliary support.

Hydrostatic bearings and hydrodynamic bearings are mainly used in the case requiring high rotation speed and high rotary precision of the spindle, such as the precision/ultra-precision spindle and the grinder spindle. For the spindle which requires higher rotational speed, the aerostatic bearing can be used. This bearing can have the rotation speed up to tens of thousands r/min and has a very high rotary precision.

3) Spindle material and heat treatment(主轴材料及热处理)

Spindle materials can be selected according to strength, stiffness, wear resistance, load characteristics, heat treatment deformation, etc. The stiffness of the spindle is related to the elastic modulus E of the material. The E value of the steel is basically the same as that of the alloy steel. Therefore, for the machine tool with common requirements, the spindle of the machine tool can be made of medium carbon steel and 45# steel with low price. The hardness of the spindle after tempering and treatment is 22—28 HRC. Alloy steel can be used when the load is high or the impact is large, or the spindle of precision machine tool is to reduce the deformation after heat treatment, or the spindle that needs to be moved in axial direction reduces its wear. The commonly used alloy steels: 40Cr, with hardness of 40—50 HRC after quenching, or 20Cr with hardness of 56—62 HRC after carburizing and hardening. Some spindle materials of high-precision machine tools are 38CrMoAl, whose hardness up to 850—1,000 HV after being nitrided(氮化).

4) Spindle lubrication and cooling(主轴润滑与冷却)

Lubrication and cooling of spindle bearings are necessary means to ensure the normal operation of the spindle. In order to minimize the impact of the thermal deformation caused by the temperature rise of the spindle components on the working precision of the machine tools, the heat of the spindle components is usually taken away by the lubricant circulation system to keep the spindle components and the box components at constant temperature; in some CNC machine tools, special cooling device is adopted to control the temperature rise of the spindle box. Some spindle bearings are lubricated with advanced grease. In order to enable the normal cooling and lubrication effect at high speed, some measures such as oil and gas lubrication, injection lubrication and raceway penetration lubrication should be adopted for some spindles.

6. Spindle quasi-stop(主轴准停)

The spindle quasi-stop is also called spindle specified position stop, whose function is to make the spindle accurately stop in the fixed circumferential position each time, so as to ensure that the end face key of the spindle can be aligned with the keyway on the tool holder when the tool is changed automatically, as shown in Figure 4-18. Meanwhile, the relative position of the tool holder and spindle is not changed when the tool is installed and the precision of the repeated installation of the tool is improved, thus the consistency of machining can be improved. In addition, for some special technological requirements, such as boring large coaxial holes in the inner wall through small holes in the front wall, or reverse chamfering, the spindle is required to achieve a quasi stop, so that the tip of the tool can stop on a fixed position and the large blade can enter the box through the front hole to bore the large holes, as shown in Figure 4-19.

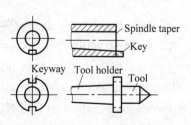

Figure 4-18 Schematic diagram of quasi stop

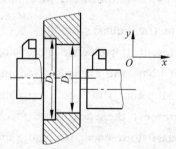

Figure 4-19 Spindle exact stop back hole

Spindle quasi stop can be divided into the mechanical and the electrical and their control processes are the same.

1) Mechanical quasistop (机械准停 jī xiè zhǔn tíng)

Figure 4-20 is a typical V-groove positioning plate quasi-stop device. V-groove positioning plate is fixed on the spindle to keep the relative position relationship between the V-groove and the end key of the spindle. After sending the quasi-stop command, a fixed low speed of the spindle is selected and the spindle begins to rotate. The non-contact switch sends out a signal to stop the main motor and disconnect the main drive chain. The spindle continues to run due to inertia, and the signal of the non-contact switch makes the positioning pin extend and contact the positioning plate at the same time. When the groove is aligned with the positioning pin, and the positioning pin is inserted into the V-groove to realize the spindle quasi-stop. The approach body of the non-contact switch should be adjusted in the circumferential direction so that the angle between the V-groove and the approach body can be guaranteed to fall into the V-groove of the positioning plate just before the spindle stops.

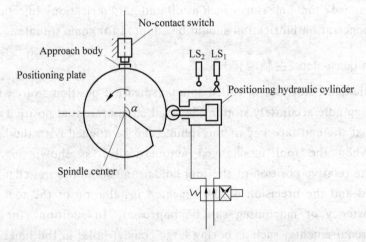

Figure 4-20 V-groove positioning plate quasi-stop device

Two-way end-face cam quasi-stop mechanism is another way, but the manufacturing of cam is more complicated. Mechanical quasi-stop methods had been outdated.

2) Electrical quasi stop (电气准停 diàn qì zhǔn tíng) control

At present, the electric quasi stop control is used in middle-and high-grade CNC systems at home and abroad. By doing so, the mechanical structure can be simplified, the quasi stop time shortened, reliability increased and the cost performance improved.

At present, there are three mechanisms of electrical quasi stop: magnetic sensor spindle quasi stop, encoder spindle quasi stop (as shown in Figure 4-21), CNC system control quasi stop.

These mechanisms are completed by CNC system. The following problems should be paid attention when adopting this kind of quasi stop control mode:

(1) The CNC system must have spindle closed-loop control function.

(2) The spindle drive device shall have the function of entering the servo state. In order to avoid the impact, the spindle drive needs the function of soft starting and so on. However, this has an adverse effect on the closed-loop control of spindle position. When the position gain is too low, the quasi stop precision and stiffness (the ability to overcome the external disturbance) cannot meet the requirements; when it is too high, severe positioning oscillation will occur. Therefore, the spindle drive must be put into the servo state, in such case, position control cannot be carried out unless the characteristic is close to that of the feed servo device.

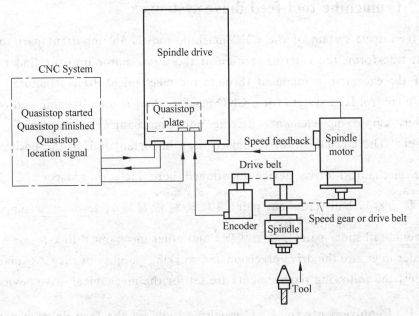

Figure 4-21 Encoder Type Spindle Quasi stop Structure

(3) Usually, for convenience, the motor shaft end encoder signal is fed back to the CNC system and the spindle drive chain precision may affect the quasi stop precision.

(4) No matter what kind of quasi stop scheme is adopted (especially for the magnetic sensor spindle quasi stop mode), the dynamic balance problem should be paid attention to when members are installed on the spindle. Because CNC machine tool spindle precision and rotational speed are very high, the requirement for dynamic balance is strict. Generally speaking, a little unbalance for the spindle under medium speed may not cause a large problem, but when the spindle rotates at high speed, this unbalance may cause vibration of the spindle. In order to meet the need of high speed of the spindle, the spindle quasi stop device of integral ring magnetic sensor has been developed in foreign countries. Because the magnetic generator is a whole ring, the dynamic balance is good.

The angle the CNC system adopted to control spindle quasi stop is set internally by the CNC system, so it is more convenient(方便的 fāng biàn de) to set the quasi stop angle. The quasi stop steps are as follows:

For example:
M03 S1000   The spindle rotates clockwise at 1,000 r/min;
M19   The spindle quasi stop is in a default position;
M19 S100   The spindle quasi stop turns to the 100° point;
S1000   The spindle rotates again at 1,000 r/min;
M19 S200   The spindle turns to the 200° point.

4.4 CNC machine tool feed drive systems

The feed drive system of the CNC machine tool is an important part for a servo system. It transforms the rotating motion of the servo motor into the linear or rotary motion of the executive component through the mechanical drive structure. A typical closed-loop control feed system for a CNC machine tool is usually composed of position comparison, amplifying element, driving unit, mechanical drive device, feedback element, etc. The mechanical drive device is an important link in the position control ring. The mechanical drive device mentioned here includes reducer(减速机构 jiǎn sù jī gòu), coupling(联轴器 lián zhóu qì), ball screw nut pair(滚珠丝杠螺母副 gǔn zhū sī gàng luó mǔ fù), lead screw support(丝杠支撑 sī gàng zhī chēng), guide rail slider pair(导轨滑块副 dǎo guǐ huá kuài fù) and other mechanical links.

In order to ensure the drive precision and working stability of the CNC machine tool feed system, the following requirements are set for the mechanical drive device.

4.4.1 Requirements for CNC machine tools on the feed drive system

1. Reduce frictional resistance(减少摩擦阻力 jiǎn shǎo mó cā zǔ lì)

In order to improve the fast response and motion precision of the feed system of CNC machine tools, the friction resistance of moving parts and the difference between dynamic and static frictions must be reduced. In order to meet the above requirements, the ball screw nut pair, the static pressure screw nut pair, the rolling guide rail, the static pressure guide rail and the plastic guide rail are widely used in the feed system of CNC machine tools. In addition to reducing friction resistance, a proper damping must be taken into account in order to ensure the stability of the system.

2. Reduce motion inertia(减少运动惯量 jiǎn shǎo yùn dòng guàn liàng)

The inertia of the moving parts has an effect on the starting and braking characteristics of the servomechanism, especially the parts and components running at high speed. Therefore, staying within the premise of satisfying the strength and stiffness of the components, the mass of the moving parts and the diameter and mass of the rotating parts should be reduced to the minimum, so as to reduce the inertia of the moving parts.

3. **High drive precision and positioning precision**(高传动精度和定位精度)

The drive precision and positioning precision of the feed drive device of CNC machine tools play a key role in the machining precision of parts, especially for the open-loop control system driven by stepper motor. Therefore, the two kinds of precision are the most important and the most characteristic indicators of CNC machine tools and also very important for point position, linear control system and contour control. In the design, by adding reducing gear into the feed drive chain to reduce the pulse equivalent (that is, the servo system receives an instruction to drive the worktable moving distance), preload the drive ball screw, eliminate the gap of the drive components, gear, worm gear, etc. We can improve the drive precision and positioning precision. It can be seen that the precision of machine tool itself, especially the precision of the servo drive chain and the servo actuator, is the main factor that affects the working precision.

4. **Wide feed rate regulation range**(调速范围宽)

The servo feed system should have a wide range of speed regulation under the condition of taking on all the workloads in order to meet the needs of various changes in working conditions, such as workpiece materials, dimensions, and cutting tools. The working feed speed ranges generally from 3—6,000 mm/min (speed range 1:2,000). In order to achieve precise positioning, the servo system has a slow approaching speed of 0.1 mm/min and in order to shorten the auxiliary time and improve the processing efficiency, the fast moving speed can be as high as 100 m/min. Such a wide range of speed is a difficult problem of the servo system design. In the CNC machine tool with multi coordinate linkage, for the synthetic velocity maintaining constant is an important condition to guarantee the surface roughness requirement. To ensure higher contour precision, the velocity of motion in each coordinate direction should also be appropriate. This is the common requirement on the CNC system and the servo feed system.

5. **Clearance-free drive**(无间隙传动)

The drive clearance of the feed system generally refers to the reverse clearance, that is, the reverse dead-time error. It exists in every drive pair of the whole drive chain and directly affects the machining precision of CNC machine tools. Therefore, the drive clearance should be eliminated as far as possible and the reverse dead-time error should be reduced. In the design, the measures such as the coupling being able to eliminate clearance and the drive pair able to eliminate clearance, etc. can be taken.

6. Quick response speed(响应速度快)

The so-called fast response refers to the speed to respond against input command signals and the speed to end the transient process, i. e. the response for tracking command signals should be fast; positioning speed and contour cutting feed speed should meet the requirements; within a specified range of speeds, the platform should be sensitive and accurate to execute commands, it should have single-step or continuous movement and should neither lose steps nor have excessive steps at run time. The response speed of the feed system not only affects the machining efficiency, but also affects the machining precision. In order to improve the rapid response of the servo feed system, the stiffness, clearance, friction and moment of inertia of the machine tool worktable and its drive mechanism should be optimized as far as possible in the design.

7. Good stability and long life(稳定性好，寿命长)

Stability is the most basic condition for the servo feed system to work normally, especially in the case of low speed feed, the stable servo feed system will not produce creeping and can adapt to the change of external load without resonance. The stability is related to the inertia, stiffness, damping and gain of the system. It is the goal of the servo system design to select appropriate parameters and achieve the best performance. The life of the so-called feed system mainly refers to the length of time for which the drive precision and positioning precision of CNC machine tools are maintained, that is, the ability of each drive component to maintain its original manufacturing precision. Therefore, the components of the feed mechanism should choose suitable materials, reasonable processing workmanship and heat treatment method. The ball screw and drive gear must have greater wear resistance and suitable lubrication to prolong their life.

8. Easy to use and maintain(易于使用和维护)

CNC machine tools are high-precision automatic control machine tools and are mainly used for the production and processing of single parts, small/ medium batches of parts, high-precision and complex parts. Therefore, the structure design of the feed system should be convenient for maintenance/care and minimize the maintenance workload, in order to improve the utilization rate of the machine tool.

4.4.2 Connection between the motor and the lead screw

The feed drive of the CNC machine has high requirements for position precision, fast response characteristic, speed regulation range and so on. There are four main kinds of motors for feed drive: the stepper, the DC servo, the AC servo and the linear motors. When different driving elements are used in the feed system of the CNC machine tool, the feed mechanism may be different. There are three main forms of connection between the motor and the lead screw, as shown in Figure 4-22.

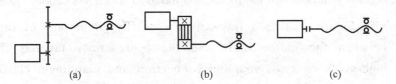

Figure 4-22 Three Forms of connection between motor and lead screw

1. Feed motion with gear drive

As shown in Figure 4-22 (a), there are two purposes of using the gear drive in the servo feed system of CNC machine tools: one is that the output of the servo motor with high-speed torque (stepper motor, DC/AC servo motor, etc.) is changed to the input of the actuator with low speed and high torque; and the other is that the rotation inertia of the ball screw and the worktable should occupy a small proportion in the system. In addition, the required motion precision should be guaranteed for the open-loop system.

Because the gear cannot achieve ideal tooth face requirement when it is manufactured, there is always a certain gear backlash, which will cause reverse loss of momentum of the feed system. For the closed-loop system, the backlash will affect its stability. Therefore, the gear drive pair usually adopts elimination measures to reduce gear backlash as far as possible, for example, using double-gear misalignment method, adjusting the center distance of gear pair with an eccentric sleeve, or using the axial gasket adjustment method to do the job.

Compared with synchronous toothed belt, the use of the gear reducer in the feed chain of the CNC machine tool is easier to produce low frequency oscillation. Therefore, a damper is often deployed in the reducer to improve the dynamic performance.

2. Feed motion driven by synchronous belt wheel(同步齿形带 tóng bù chǐ xíng dài)

As shown in Figure 4-22 (b), the mechanical structure of this form of connection is relatively simple. Synchronous belt drive is a new type of belt drive. It uses the meshing in turn between the tooth shape of the toothed belt and the gear tooth of the belt wheel to transmit motion and power, thus having the advantages of the belt, the gear and the chain drives; in addition, it is characterized by no relative slippage, very accurate average drive, high drive precision as well as high strength, small thickness and light weight of the toothed belt. Therefore, it can be used for high-speed drive. The gear belt does not need special tension, so the load on the shaft and bearing is small, the drive efficiency is also high and it is widely used in places with low torque characteristics. However, there are strict requirements for center distance at the time of installation and the manufacturing process of synchronous belt, and the belt wheel is complicated.

3. Direct connection between the motor and the lead screw via coupler(联轴器)

As shown in Figure 4-22 (c), this structure usually adopts any of the following connections between the motor shaft and the lead screw: tapered ring keyless connection, high-precision cross coupling connection and diaphragm elastic coupling connection, so that the feed drive system can have higher drive precision and drive stiffness and the mechanical structure can be greatly simplified. This kind of connection is widely used in the feed movement of the machining center and the CNC machine tool with high precision.

When the machine is running, the two shafts cannot be separated. The two shafts can be separated only after the machine stops and the coupler is removed. It can be seen that the function of the coupler is the component to connect axially the two shafts and transfer the torque and the moving parts and it has the ability to compensate the biaxial offset to a certain extent. In order to reduce the vibration of the mechanical drive system and reduce the load of shock peak, the coupling should also have certain damping and shock absorption ability. The coupler sometimes also concurrently has the function of overload safety protection.

At present, there are many types of couplers, including hydraulic, electromagnetic and mechanical types. The mechanical coupler is the most widely used and it transfers torque by means of the interaction force between mechanical components. Mechanical couplers can be divided into the rigid and the flexible couplers.

Rigid coupler(刚性连杆): The rigid coupler does not have the ability to compensate the relative deviation of the axis of two shafts connected, nor does it have the capability of cushioning and damping, but its structure is simple and its price is low. The rigid coupler can be selected only if both the load and the rotational speed are stable and the relative deviation of the axis of the two shafts connected can be guaranteed to be minimal. Figure 4-23 shows the sleeve coupler while Figure 4-24 shows the flange coupler.

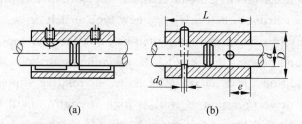

Figure 4-23 Sleeve coupler

Flexible coupler(挠性连杆): The flexible coupler has the ability to compensate the relative deviation of the axis of the two shafts connected and the maximum value varies with different models. Figure 4-25 shows the miniature diaphragm coupler.

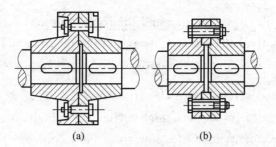

Figure 4-24 Flange coupler

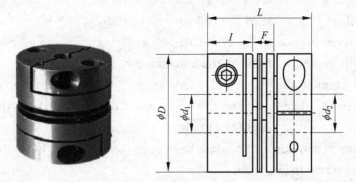

Figure 4-25 Miniature diaphragm

As the feed speed of the CNC machine tool is fast, for example fast forward and backward speeds may be as high as 20—30 m/min sometimes, in the whole machining process, clockwise and counterclockwise conversion is frequent. The coupler is subject to large impact and thus easy to cause coupler loosening and torsion. As the working time increases, its loosening and torsion will increase. In actual processing, the main situation is that the movement of the coupler in each direction is normal, the feedback of the encoder is also normal and the system has no alarm, but the value of motion cannot always accord with the value of the command and the error value of machining is increasing, even resulting in the scrapping of processed parts. In this case, check the coupler promptly.

4.4.3 Ball screw nut pairs

1. Profile of the ball screw nut pair

At present, there are two main drive modes of feed motion widely used: one is the feed motion by the indirect drive of the rotary servo motor through the ball screw nut pair, the other is the feed motion by the direct drive of the <u>linear motor</u>(直线电动机). The latter is mainly used in high-speed machining. The ball lead screw nut pair ("ball screw pair" for short) is a kind of lead screw pair which is equipped with balls between the screw and the nut as the intermediate drive element and it is a drive device which can be converted between the linear motion and the rotary motion(直线运动和转动). As the lead screw rotates, the balls rotate at its axis in the raceway and also circularly rotate along the raceway, thereby forcing the nut (or lead screw) to move axially. Compared with the traditional lead screw, the ball screw nut pair(滚珠丝杠螺母副) has the advantages of high drive precision, high efficiency, high stiffness, possible pre-tightening, steady movement, long life, low noise and so on. The details are as follows:

(1) Because the rolling friction is used to replace the sliding friction, the friction loss and the starting torque are smaller, and the drive efficiency is as high as 85%—98%, which is 3—4 times that of the ordinary sliding screw pair.

(2) The axial clearance can be eliminated by applying the pretension force, thus the positioning precision and stiffness of the machine tool can be improved. The ball screw pair takes into account the impact of tooth shape precision, material characteristic, surface hardness and other factors on the lead screw life. And it adopts rolling friction, so it not only has a long service life, but also is predictable.

(3) In recent years, some companies have improved the backflow unit to reduce the noise of the lead screw by 5—7 dB.

(4) The static and dynamic friction coefficients of the ball screw are virtually not different. It can eliminate the reverse clearance and apply preload, which helps to improve the positioning precision and stiffness.

(5) The manufacturing cost of the ball screw is high, for example, the screw groove raceway in the screw nut usually requires grinding into shape and the process is complicated.

(6) Since the friction angle is less than 1° and cannot be self-locked, especially the vertical lead screw, the main shaft box can be easily caused to slide due to self-weight and inertia or sudden power failure, so in this case a braking device is required.

(7) At present, many manufacturers at home and abroad have started the standardized, serialized, modularized and specialized production.

2. Circulation of the ball screw pair

There are two common circulation modes: external circulation and internal circulation (外循环和内循环).

(1) External circulation. In this case, the ball is sometimes disconnected from the lead screw during circulation, as shown in Figure 4-26. The external circulation is further divided into embedded cannula type, cannula penetrating type, end block type and end cover type. The cannula mode has a spiral cannula on the outer circle of the nut. The two ends of the cannula are inserted into the holes in the top and the end of the nut to guide the balls to form a circulation chain through the cannula. There are two kinds of structures

for external circulation: single-row and double-row. The former mode is simple in structure, good in technology, high in carrying capacity, but large in radial size.

(2) Internal circulation. In this case, the ball is always in contact with the lead screw during the circulation, as shown in Figure 4-27. This structure crosses two adjacent raceways with the reverser, and the balls enter the adjacent raceway through the reverser from the thread raceway, forming a closed-loop circuit. Generally, a nut is equipped with 2 to 4 reversers, which are evenly distributed around the nut. As one row has one ring of balls only, there are a small number of working balls, the balls are characterized by good smoothness, small friction and high efficiency. The radial size of nuts in this structure is small, but it is difficult to manufacture. There is also the end cover type circulation, which is similar to the external circulation. The main difference in-between is that the spiral groove is milled out in the outer circle of the nut and the two ends of the groove are drilled through holes, which are tangent with the threaded raceway of the nut to form the ball return channel. To prevent the ball from falling off, the spiral groove is covered with a steel sleeve. The length of the nut can be shortened under the same dynamic load because it is covered with steel balls on the raceway of the nut.

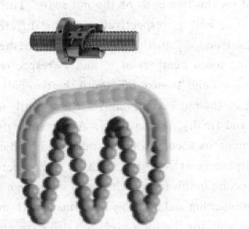

Figure 4-26　External circulation reflux　　　Figure 4-27　Internal circulation reflux

3. Pre-tightening of the ball screw pair

The purpose of the pre-tightening of the ball screw pair is to eliminate the clearance between the screw and the nut and apply the pre-tightening force to ensure the precision and axial stiffness of the ball screw reverse drive.

The pre-tightening methods used in the feed system of CNC machine tools include grinding the thickness of gasket, clearance elimination of double locknuts and adjustment of tooth difference. The double-nut structure is widely used to eliminate the clearance.

(1) Grinding the thickness of gasket to eliminate clearance, as shown in Figure 4-28. By adjusting the thickness of the gasket, the two nuts respectively on the left and right may produce axial displacement to eliminate the clearance between the lead screw nuts and

produce pre-tightening force. This method has simple structure but is inconvenient to adjust. In addition, when the raceway worn out, it could not eliminate clearance and would need pre-tightening at every time.

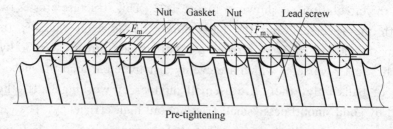

Figure 4-28 Backlash of gasket structure

(2) Eliminating clearance with double locknuts, as shown in Figure 4-29. Two locknuts are used to adjust the pre-tightening force of the lead screw nuts. This method is convenient and easy but cannot adjust the pre-tightening force accurately.

(3) Eliminating clearance with tooth difference, as shown in Figure 4-30. On the flange of the two nuts, the cylindrical outer gears are respectively meshed with the inner rings which are respectively fastened on the two ends of the nut seats. The number of teeth of the cylindrical outer gears is z_1 and z_2 respectively and the difference is one tooth, i.e. $z_2 = z_1 + 1$. Load the nuts 1 and 2 into the nut seat 5 and meshed with the inner gears 3 and 4 with corresponding tooth numbers of z_1 and z_2 respectively. When adjusting, firstly remove the inner gears 3 and 4, turn one tooth the two ball nuts in the same direction toward the nut seat (i.e. the nut 1 axially moves P_n/z_1 while nut 2 moves P_n/z_2), then insert the inner gears and finally tighten them. As a result, the two nuts will produce relative axial displacement to accomplish the purpose of adjusting axial clearance and pre-tightening. The displacement is $s = (1/z_1 - 1/z_2)P_n$, where P_n is the lead distance of the lead screw. The structure in this form is characterized by high complexity, large size as well as poor machining workmanship and assembly performance. However, it can achieve accurate adjustment and is suitable for the high-precision drive structure.

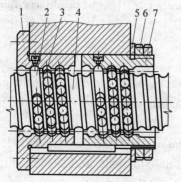

Figure 4-29 Double-locknut clearance elimination

1,7—Nut; 2—Reverser; 3—Ball;
4—Lead screw; 5—Gasket; 6—Round nut

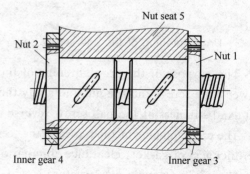

Figure 4-30 Tooth difference clearance

4. Application of the ball screw pair

1) Support of the ball screw pair

The installation mode has great impact on bearing capacity, stiffness and maximum rotation speed of the ball screw. The ball screw nut pair should meet the following requirements when installation.

(1) The ball screw nut pair cannot have axial play with the relative worktable;

(2) The center of the nut seat hole should be concentric with the axial line of the lead screw installation;

(3) The ball screw nut pair should be parallel to the central line of the corresponding guide rail;

(4) Easy clearance adjustment, pre-tightening and pre-stretching.

There are the following four common installation modes.

(1) Fixed-free (G-Z mode) (一端固定一端自由 yì duān gù dìng yì duān zì yóu). As shown in Figure 4-31(a), thrust angular contact ball bearings or needle/thrust cylindrical roller bearings which can withstand bidirectional and radial loads are installed only at one end and are pre-tightened in an axial direction while the other end is completely free and has no support. This support structure is simple, but the bearing capacity is small. It is suitable for short axial lead screw with low speed, medium precision, short lead screw length and short stroke.

(2) Fixed-supported (G-J mode) (一端固定一端支撑 yì duān gù dìng yì duān zhī chēng). As shown in Figure 4-31(b), one end of the lead screw is fixed and the other end is supported. The fixed end bears axial force and radial force at the same time while the supported end bears only radial force and can have slight axial floating. This support structure can reduce or avoid the bending due to the self-weight of the lead screw and the thermal deformation of the lead screw may freely extend to one end. It is suitable for medium rotation speed and high precision.

(3) Fixed-Fixed Dual-thrust (G-G mode) (两端固定 liǎng duān gù dìng). As shown in Figure 4-31(c), both ends of the lead screw are fixed. The thrust angular contact ball bearings or the needle/thrust cylindrical roller bearings, which bear bidirectional and radial loads, are installed at both ends and preloaded to improve the support stiffness of the lead screw and partially compensate the thermal deformation of the lead screw. It is suitable for high rotation speed and high precision.

(4) Supported on both ends (J-J) (两端支撑 liǎng duān zhī chēng). As shown in Figure 4-31 (d), both ends of the lead screw are supported. The support in this form is simple, but the supported ends bear only radial force, the lead screw will extend after thermal deformation, thus affecting the machining precision. It is suitable only for medium rotation speed and medium precision.

The ball screw pair, as a precise, efficient, sensitive drive element, in addition to

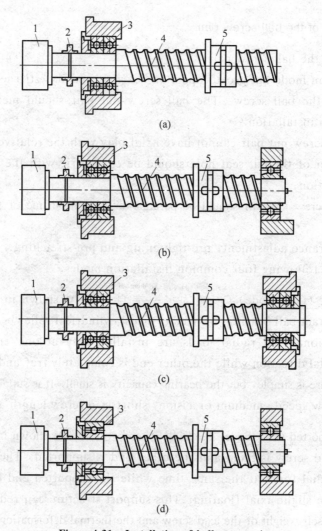

Figure 4-31 Installation of ball screw pair
1—Motor; 2—Flexible coupler; 3—Bearing; 4—Ball screw; 5—Ball screw nut

the use of high-precision lead screw, nut and ball, should select the bearing with axial stiffness, small friction torque and high operating precision. The support for the ball screw commonly uses bidirectional thrust angular contact ball bearing, tapered roller bearing, needle and thrust roller composite bearing, deep groove ball bearing, thrust ball bearing, etc. At present, the single-row thrust angular contact ball bearing with 60° contact angle is most widely used as the support of the ball screw and its axial stiffness is more than twice as high as that of the general angular contact ball bearing and the thickness of the inner and outer rings has been selected when the product is out of the factory. As long as the nut and the end cover are used to press the inner ring and the outer ring, the factory pre-tightening force can be obtained and the operation is convenient.

When purchasing, the end of the ball screw should be ordered from the manufacturer according to the design requirements and the shaft shoulder or thread can be processed.

2) Braking mode (制动方式 zhì dòng fāng shì)

The drive efficiency of the ball screw pair is high, but it cannot be locked automatically. A brake device is required for vertical drive or high speed and large inertia. At present, there are two common braking modes, i.e. the mechanical and the electrical. Electrical braking applies an electromagnetic brake and this type of brake is within the motor. Figure 4-32 shows Fanuc servo motor with the brake. When the machine tool is working, under the action of the electromagnetic force of the brake coil 7, the gear 8 is separated from the inner gear 9, the spring is compressed and when the machine stops or goes out of power, the electromagnet loses power, and the gears 8 and 9 mesh with each other under the action of the spring recovery force; gear 9 is integrated with the end cover of the motor, so the lead screw connected with the motor shaft is braked. The electromagnetic brake is mounted in the motor housing and forms an integrated structure with the motor.

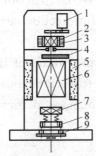

Figure 4-32 Ball screw electrical braking
1—Rotary transformer; 2—Speed generator rotor; 3—Speed generator stator;
4—Brush; 5—Permanent magnet; 6—Servo motor rotor;
7—Electromagnetic coil; 8—External gear; 9—Internal gear

The other braking modes are as follows:

(1) Use the braking motor with the function of drive braking.

(2) High reduction ratio system with low reverse efficiency in the drive chain, such as gear and worm reducer. This method depends on friction to achieve the purpose of braking, so it is not economical.

(3) Use overrunning clutch.

4.4.4 Guide rail slider pairs

1. Requirements of CNC machine tools on the guide rail slider pair

The guide rail is mainly used to support and guide the moving parts to move along a certain track, so as to ensure the relative position and relative position precision of each component. To a great extent, the guide rail determines the stiffness, precision and precision of the CNC machine tool. Therefore, the machine tool requires the following characteristics of the guide rail.

1) High guiding precision

Guiding precision refers to the precision of the mutual position of the track of the moving parts of the machine tool along the guide rail and the relevant reference plane of the machine tool. The main factors affecting guiding precision are manufacturing precision, structure form, assembly quality, the stiffness and thermal deformation of guide rail and its supporting parts.

2) Good wear resistance

Wear resistance directly affects the precision retention of machine tools. The long-term motion of the moving guide rail surface along the supporting surface may lead to uneven wear of the guide rail, thus affecting the machining precision of the machine tool. When choosing a guide rail, we must consider its wear resistance and minimize its unevenness. The guide rail should be able to recover automatically. The factors affecting wear resistance include: guide rail material, friction property, treatment and processing method, stress condition, sliding and protection during work.

3) Large stiffness

The deformation of the guide rail will affect the relative position and the guiding precision of the moving parts. Therefore, the guide rail should have enough rigidity to ensure that it does not produce too much deformation after force is applied, thus affecting the machining precision.

4) Good friction property

The friction coefficient of the guide rail should be small to reduce the friction resistance and the thermal deformation of the guide rail and the coefficients of static and static friction should be as close as possible, so that the movement can be stable and there is no crawling at low speed.

2. Classification of guide rail pairs

According to the friction properties of the contact surface, the guide rail pairs can be divided into the sliding, the hydrostatic and the rolling types.

1) Sliding guide rail

The sliding guide rail is divided further into two categories: the metal-to-metal and the metal-to-plastic. The static friction coefficient of metal-to-metal type is large and the dynamic friction coefficient varies with the velocity and it is easy to creep at low speed. Compared with the general guide rail, the plastic guide rail has the characteristics of stable chemical composition of plastic, small friction coefficient, good wear resistance, strong corrosion resistance, good vibration absorption, small specific gravity, simple processing and forming and can work under any liquid or non-lubricated condition. It is applied in CNC machine tools. There are two categories of plastic guide rails, i.e. the PTFE guide rail soft tape and epoxy wear-resistant guide rail coating. In usage, the former uses the pasting method, so it is commonly called "Plastic Guide Rail". The latter coated guide rail is painted or scraped or injected into with pasty plastic, so it is commonly called "Injection Guide Rail". The shortcomings of the plastic guide rail include poor heat resistance, low thermal conductivity, larger coefficient of thermal expansion than metal, easy deformation under external force, poor rigidity, large hygroscopicity and dimensional stability affected.

2) Hydrostatic guide rail

Hydrostatic pressure guide rail means that the pressure oil enters the two opposite moving guide rail surfaces through the throttle and then the oil film is formed to separate the two guide rail surfaces, which ensures the guide faces to work in the liquid friction state. In operation, the oil pressure on the oil chamber on the rail surface is automatically adjusted with the change of the applied load. This guide rail has good friction characteristics and its friction coefficient is linear with the velocity, but the change is very small and the starting friction coefficient can be as small as 0.0005. It has strong vibration absorption, steady motion and no crawling. And it is applied to large and heavy CNC machine tools with high precision and high efficiency. Its structure can be divided into two types: the open and the closed. For the latter, there is an oil cavity on the surface of each direction of the guide rail, so it has the ability to bear the load in all directions and can maintain a good balance. In addition to liquid hydrostatic guide rail, there is the gas hydrostatic guide rail, also known as air cushion guide rail. The friction coefficient is smaller than that of the hydrostatic guide rail.

3) Rolling guide rail

The rolling guide rail places rolling objects such as the ball, roller and needle between the guide surfaces so that the friction in-between is rolling friction. This guide rail has the advantages of high motion sensitivity, high positioning accuracy, good precision retention and convenient maintenance. Figure 4-33 shows the shape of the linear rolling guide rail.

Figure 4-33 Linear rolling guide

The rolling guide pair can be divided into two categories according to the structural styles, i.e. the rolling guide and the roller guide rail block. The rolling guide rail is mostly used in medium and small CNC machine tools. The roller guide rail block, with rollers as rolling objects, is used in conjunction with the guide rail of the machine tool bed. It is not limited by the length of travel, and has strong bearing capacity as well as stiffness, but its friction coefficient is a little larger; it is commonly used in heavy-duty machine tools. At present, the rolling guide pair has become a standard part by the special manufacturer. The specifications are set by the manufacturer, but they all work in the same way.

The rolling guide rail pair can be divided into the rolling linear guide, and the rolling arc guide according to shape. The rolling arc guide rail pair can realize the arc or circular motion of any diameter, overcoming the dimension limitation caused by the machining of the bearing or rolling support and so on. Theoretically, the larger the diameter of the arc guide rail pair, the more convenient to design, manufacture, install, maintain, etc.

According to the relationship between the guide rail and the slider, the rolling linear guide pair is divided into the integral guide pair and the separate guide pair. For the latter, the preload between the guide rail and the slider can be adjusted arbitrarily in practice to improve the rigidity of the system or the stability of the motion and the height of the guide rail pair is very low. It is possible to achieve precise linear steering motion in a very small space.

In terms of precision, the foreign brands of rolling linear guide rail pairs can be classified into three grades, i.e. P, H and N and whereas the domestic brands can be divided into three grades, i.e. 3, 4 and 5. Among them, grade N or grade 5 has the lowest precision and is relatively cheaper. The manufacturer also sets the guide rail types according to the width of the rail cross section, if the width is 16 mm, it is 16♯ guide rail. When purchasing a rolling linear guide rail pair, we should note whether the slider has a flange.

According to whether there is a ball holder in the guide rail pairs, rolling linear guide rail pairs can be divided into the normal type and the low-noise type. Because the ball holder makes the oil film contact between the ball and the retainer, so that the friction between the balls can be avoided and the heat of the guide rail pair in operation can be greatly reduced, thus the stable operation can be realized and the high-speed and high-precision movement of the guide rail pair can be realized.

3. Structure of the rolling guide rail

The rolling linear guide pair (gǔn dòng dǎo guǐ fù 滚动导轨副), as shown in Figure 4-34, is a linear guide rail pair with a ball holding device and equal load in four directions forming a very common guide rail pair. In the Figure, 1 is a supporting guide rail with a slider 8 installed on it. Four groups of balls are arranged in the slider 8 and roll in the linear raceway between the supporting rail and the slider. When the balls 3 roll to the end of the slider, they go back to the other end through the baffle 6 and circulate like this. 4 is the cage and 2 is the side gasket. The gasket 5 is used to prevent dust from entering the guide rail. 9 is the lubricant nozzle through which lubricant can be injected. As shown in Figure 4-35, the balls 3 are separated by the retainer 7 and the balls do not collide with each other, thus reducing the noise generated in the motion of the guide rail pair, so the guide rail is called the low-noise rolling linear guide rail. Four groups of balls and raceways are equivalent to four linear angular contact bearings (jiǎo jiē chù qiú zhóu chéng 角接触球轴承), the contact angle is 45°. The four directions have the same bearing capacity.

In addition to the ball rolling guide rail described above, there is also the roller

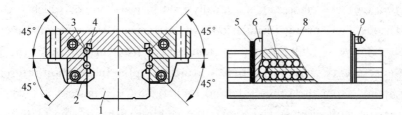

Figure 4-34　Linear rolling guide

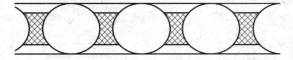

Figure 4-35　Retainer

linear guide as shown in Figure 4-36 (a). Because the contact area between the roller and the guide rail is larger than that of balls, the guide rail has larger load capacity. Load capacity can also be improved by increasing the number of raceways, as the 6-raceway rolling guide rail shown in Figure 4-36 (b).

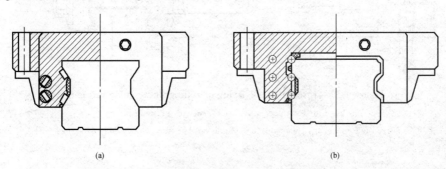

Figure 4-36　Other types of guide

(a) Roller linear guide rail; (b) 6-raceway rolling guide rail

ān zhuāng yǔ yù jǐn
4. Installation and pre-tightening（安 装 与 预 紧）of rolling guide rails

The mounting and fixing methods of linear rolling guide rails are mainly screw fixing, pressing plate fixing, dowel pin fixing and oblique wedge fixing, as shown in Figure 4-37. The installation forms of linear rolling guide rails can be horizontal, vertical or inclined, it can be installed in parallel with two or more rails, or two or more short guide rails can be connected to meet the needs of various travels and applications. The linear rolling guide pair can simplify the design, manufacture and assembly of the guide rail of the machine tool. The precision of base surface installation of the rolling guide rail is not too high. Usually only fine milling or fine planning is required. Because of the homogenization effect of the linear rolling guide rail, the error of installing the base surface will not completely reflect the motion of the sliding seat and usually the motion error of the slide seat is about 1/3 of the base plane error.

In practical use, two guide rails are usually used in pair, one of which is the reference guide rail. Through its correct installation, we can ensure the correct direction of the moving parts relative to the supporting elements. During installation, the positioning surface of the reference guide rail is tightly attached to the installation reference surface and it is tightened from the other side and then fixed. In Figure 4-37, (a) Shows the jacking and fixing with screws. (b) Shows the jacking with pressing plate or the use of tightening screws in the pressing plate. (c) The side surface of the guide rail is assembled and the workmanship is very good. (d) Shows the jacking with the wedge.

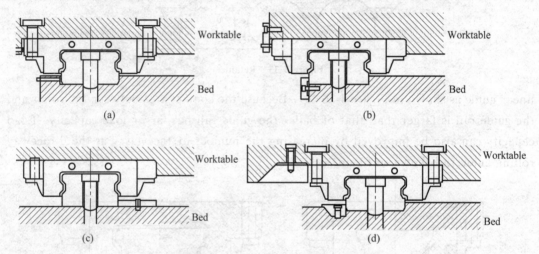

Figure 4-37　Installation & fixing of rolling linear guide rail pair
(a) Fixed with screws; (b) Fixed with pressing plate;
(c) Guide rail bar with/dowel; (d) Guide rail bar fixed with wedge & screws

4.5　Self-made one-dimensional feed mechanism

For the self-made one-dimensional feed mechanism(自制一维进给机构), see Figure 4-38.

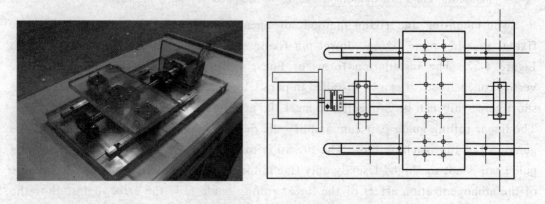

Figure 4-38　Material product

The chosen parameters are as follows:

1) Structural form and main parameters of the ball screw pair chosen

Choose nominal diameter: 20 mm; nominal lead: 4 mm; diameter of steel ball: 2.381 mm; left screw thread; number of circles: 1; number of rows: 3; thread rising angle: 3°38′; specification code: FFZL2004LH-3.

2) Bearing type

Choose a support with one-end fixed and the other flexible; choose back-to-back 60° thrust angle contact ball bearing for the fixed end and deep groove ball bearing for the flexible end. Choose 7001ADF double-row thrust angular contact ball bearing for the fixed end and 6000ZZ deep groove ball bearing for the supporting end.

Guide rail model: SBG15SL.

3) Coupler type

SRBM-32C flexible coupler.

Chapter 5　Intelligent manufacturing

5.1　The development and application of intelligent manufacturing

U.S. industrial Internet, the German industrial 4.0 and Made in China 2025, are typical representatives of the national strategies of the major powers. Although the expression may vary in these countries, the core is always the same, namely, the intelligent manufacturing. All countries are building their own intelligent manufacturing system, which lays the foundation for the technical system, standard system, and for the industrial system. In fact, the new wave of industrial revolution is essentially the standard competition in the future.

5.1.1　American industrial Internet

The United States is the world leader in both the theory and the application of intelligent manufacturing technology, basically, the intelligent technology, such as artificial intelligence(AI, 人工智能), cybernetics, internet of things(物联网), mostly originated in the U.S. From the early CNC machine tools, integrated circuits(集成电路), PLC, to today's smart phones, driverless cars and a variety of advanced sensors, nearly all of them are developed in American university laboratories and corporate R&D centers. In 2010, President Obama signed the "American Manufacturing Promotion Act", put forward the "reindustrialization" (i.e. manufacturing return) strategy which underlines the importance of the rapid development of AI, robotics and digital manufacturing for the United States to reconstruct the competition in the manufacturing industry. In March 2012, National Network of Manufacturing Innovation (NNMI) was proposed, and nine central innovation projects were listed among the four key areas of manufacturing, meanwhile, the collaboration between universities and enterprises would be enhanced by means of establishing research centers. In this process, the United States Industry Alliance took the lead in the formation of the industry, which has been an important driving force of the Internet revolution. In 2012 General Electric (GE) proposed the concept of the industrial Internet. This network can connect the machine to sensors, control to software applications, so as to improve production efficiency and reduce resource consumption. In 2014, GE and four IT companies—IBM, CISCO, Intel and AT&T formed the industrial Internet alliance. The alliance "opened up" its membership to establish a community with "no scientific barriers" and to further promote the integration of the real physical world and the big data of the digital world. The industrial Internet has transformed the manufacturing industry. Tesla Motors and Google automatic driving cars will become Internet terminal and function exactly the same as the mobile phone or computer. The information technology has boosted the development of the manufacturing industry(制造业) and adjustment of business strategy. The cross-industry collaboration continues to be strengthened so as to enhance the intelligence and networking of the manufacturing industry.

American industrial Internet connects people, data and machines, and the formation of an "open, global" industrial network contributes to the subversion of software, network, big data services for industrial areas and improvement of value creation of the whole industry with the aid of Internet and data. GE, for example, is the world's leading manufacturer of aircraft engines. The various sensors of GE aircraft engines are able to collect all kinds of data during the flight, and the data are transmitted(传输) to the ground and analyzed by the intelligent software system, therefore, the aircraft operating conditions can be accurately detected, and even a fault in the plane can be predicted so that the timely preventive maintenance can be made to improve the safety as well as prolong the life of the engine(发动机). In fact, throughout the product life cycle of GE aircraft engine, the sales of physical products only account for only 30% in the value creation, whereas the service and maintenance account for 70% of total revenue. The major source of profits comes from the service instead of the product itself.

5.1.2 Germany industry 4.0

Germany, as an industrial powerhouse, is taking the lead in the core competition, but with the integration of the new-generation Internet technology and industry in depth, gradual development of intelligent products, equipment, technology and services, all walks of life in Germany is deeply worried about whether they can keep up with the pace of development, especially when the Internet giants of United States involved in the industrial field. Therefore, in 2011, the German Research Center for AI, President and CEO of Wolfgang professor Wahlster put forward the "industry 4.0" concept, which aims at the seamless docking of the real industrial production and the virtual digital world through the facilitation of the Internet.

In April 2015, in the report of the "the industry 4.0(工业 4.0) strategic plan", Germany made a strict definition of it, which was referred to as the fourth industrial revolution, it means that the life cycle of product would have further control for the value creation chain organization from the innovative ideas and orders, to R&D, production, delivery and even the waste recycling and relative service industry. In each stage, the needs of customers could be satisfied in a more personalized way. All entities involved in the creation of the value of the formation of the network, access to data from time to time to create the maximum value of the ability to achieve real-time sharing of all relevant information. On this basis, through the connection of people, objects and systems, the dynamic establishment of enterprise value network, real-time optimization (实时优化) and self-organization could be achieved. And the cost, efficiency and energy consumption(能耗) could be optimized according to different standards. In short, the fourth industrial revolution in Germany is characterized by intelligent value creation as the core of the industrial revolution(工业革命), to solve the problems such as the customized production technology, the complex process management, the analysis of big data, the optimization of decision process and the implementation of action.

The key points of Germany industrial 4.0 strategy framework(框架) are shown in Figure 5-1, namely 1 network (cyber physical system (CPS) network), 4 themes (intelligent production, smart factories, intelligent logistics, intelligent service), 3 integrations (vertical integration, horizontal integration, end-to-end integration and technology) and 8 objects (standardization and reference architecture, management system, broadband infrastructure, industrial complex, safety and security work of the organization and design, training and reeducation, regulatory framework, resource utilization efficiency).

As the world's leading supplier of equipment manufacturing, based on its excellent equipment manufacturing tradition and advanced information infrastructure, German has placed a special premium on developing intelligent equipment. At present, SIEMENS, BOSCH, SAP and other companies have been chosen as the core to develop Germany industry 4.0, due to their superiority in industrial automation(自动化), equipment development, and assets management. Germany industry 4.0 concentrates more on the intermediate and microscale, stressing the "hard" manufacturing, whereas the U.S. industrial Internet put more emphasis on the macroscale, highlighting "soft" service, although their logical paths(逻辑路径), the chief goals are consistent.

5.1.3 Made in China 2025

With the continuous development of intelligent manufacturing technology in the world, many countries are readjusting the layout of the manufacturing industry. China's overall internal and external economic environment has changed, the advantage in low cost, late development, and technology acquisition in the manufacturing sector is

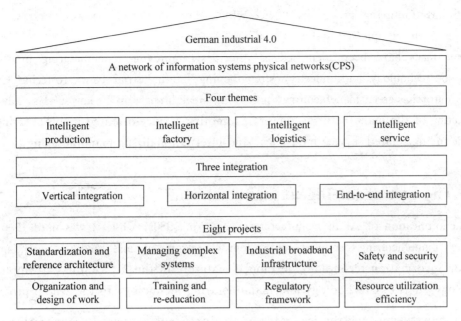

Figure 5-1　The strategic framework of Germany industry 4.0

gradually disappearing. The rapid development of intelligent manufacturing will accelerate the reconstruction of the pattern of global manufacturing, Chinese manufacturing industry has been faced with the double competition(双重竞争 shuāng chóng jìng zhēng) from both the developed and the developing countries, which requires the Chinese manufacturing industry must take an international perspective, meet the challenge and become a major manufacturing power in the global competition.

In 2015, China formally promulgated(颁布 bǎn bù) the "Made in China 2025", as a guideline for a new round of industrial revolution, which underlined the integration of industrialization and information, striving for the breakthrough in the 10 priority sectors: ① New advanced information technology; ②Automated machine tools & robotics; ③Aerospace and aeronautical equipment; ④ Maritime equipment and high-tech shipping; ⑤ Modern rail transport equipment; ⑥New-energy vehicles and equipment; ⑦Power equipment; ⑧Agricultural equipment; ⑨New materials; and ⑩ Biomedicine and advanced medical products.

The guidelines of Made in China 2025 are innovation-driven, giving priority to quality, green development, optimizing structure, and talent-oriented. And the principles are market-oriented and government-guided; based on the present and having a long-term perspective; comprehensively pressing forward and making breakthroughs in key areas; independent development and win-win cooperation. The goal is to be achieved through a "three-step" strategy, and each step will require about ten years. In the first step, by 2025 China will be ranked among the manufacturing powers; in the second step, by 2035 China's manufacturing sector will reach a generally moderate level among the manufacturing powers; the third step will mean transforming China into a leading manufacturing power by 2049, which will be the 100th anniversary of the founding of the People's Republic of China.

Nine tasks have been identified as priorities: improving manufacturing innovation, integrating technology and industry, strengthening the industrial base, fostering Chinese brands, enforcing green manufacturing, promoting breakthroughs in 10 key sectors, advancing restructuring of the manufacturing sector, promoting service-oriented manufacturing and manufacturing-related service industries, and internationalizing manufacturing.

5.2 The core of intelligent manufacturing

The application of AI in manufacturing began in 1980s, while it was until 1990s that intelligent manufacturing technology and system were proposed, and intelligent manufacturing based on information technology developed maturely in the 21st century. It applies AI technology, network technology and manufacturing technology to the whole process of product management and service, and can analyze(分析), reason and perceive in the manufacturing process of the product to meet the dynamic needs of the product. At the same time, it also changes the manufacturing mode of production, man-machine relationship and business models, which therefore, is not a simple intelligent manufacturing technology breakthrough, it is not a simple transformation of traditional industries, but an innovation and integration of communication technology(通信技术) and manufacturing. Intelligent manufacturing covers 5 aspects: intelligent products, production, equipment, management and services, which aims at more rational(理性的) and intelligent use of equipment and maximization value through the intelligent operation and maintenance(维护).

The three traditional industrial revolutions have centered on the technological upgrading of the five elements in the manufacturing system: material, equipment, process, measurement and maintenance. Whether the enhance of the precision of equipment and the level of automation, or the use of statistical quality management, or improvement(改进) of equipment monitoring availability or the high industrial production efficiency by the lean manufacturing system, these activities are always focusing on people's experience. The logic of the operation always follows the same pattern: finding the problem, analyzing the problem according to the experience, adjusting the 5 elements according to the experience, and solving the problem.

What distinguished the intelligent manufacturing different from the traditional manufacturing system lies in sixth elements, that is, modeling, and the other 5 elements are to be driven by the sixth elements, so as to solve and avoid the problems of manufacturing systems. Therefore, operation logic(运行逻辑) of intelligent manufacturing system is:

problems arising, model analysis, adjustment of five elements, problem solving, model accumulation of experience, analysis of root problems, model of 5 elements, problem avoiding. Therefore, whether or not the manufacturing system can be called "intelligence" will depend on the following two characteristics:

(1) The ability to learn from the experience of humans, replacing people to analyze problems and make decisions;

(2) The ability to continually accumulate experience from the new problems, and to avoid the recurrence(复现) of similar problems.

The key problem to be solved is to model the activities of the five elements of the manufacturing system. Problems in the manufacturing system can be divided into "the visible" and "the invisible", the intelligent manufacturing is based on a comprehensive understanding of "the visible" and "the invisible" to avoid the problems on the basis of the worry-free manufacturing environment. In order to accumulate experience and knowledge in the process of solving the problems of the visible, the main carrier of American is the accumulation of data, the Japanese carrier, Germany carrier of the equipment, which respectively formed their own characteristics and advantages of manufacturing system. With the large amount of product lifecycle(产品生命周期) data acquisition, especially the development of the advanced analysis technology of the datum collection, machine learning and simulation modeling, people can therefore obtain previously invisible knowledge from data, and use it to manage and solve previously invisible problems.

5.3　Technical support for intelligent manufacturing

Intelligent manufacturing technology can be classified into 9 categories: industrial networking, cloud computing, big data industry, industrial robots, 3D printing, industrial automation, working knowledge of network security, virtual reality (VR) and AI. Among which, the industrial networking, cloud computing and big data industry are three underlying infrastructures in the Internet era, industrial robots and 3D printing (3D 打印) are the two major hardware technologies. Knowledge work automation and industrial network security mainly provide the software support, VR and AI are two ultimate technologies for the future, as shown in Figure 5-2.

1. Industrial Networking

Industrial networking is the use of local network or Internet communication technology, having the sensors, controllers, machines, personnel and goods linked together through a new way to establish connection between people and things, and of all kinds of objects, so that the information technology, remote control and intelligent

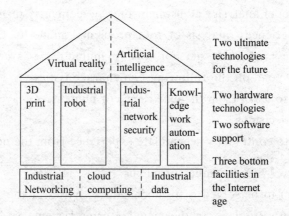

Figure 5-2　Technical support of intelligent manufacturing

network can be realized to maximize the efficiency of the machines and the throughput of the work. Distributed physical information system (CPIS) is the core technology of industrial IOT intelligent system. What is CPIS? It is a multidimensional intelligence system(多维智能系统)(duō wéi zhì néng xì tǒng), based on big data, network and massive computing. The key technologies in the intelligent sensing, analyzing, mining, evaluation, prediction, optimization, collaborative technology fields will be used to achieve organic integration and cooperation between calculation, communication and control, especially the integration of the VR and the physical world regarding the objects mechanism(机制)(jī zhì), environment, and community.

The establishment of CPIS will enable the physical devices (sensors) to be connected the Internet, which can realize functions such as computation, communication, precision control, remote coordination(远程协作)(yuǎn chéng xié zuò), autonomy, data acquisition and other functions, so as to achieve the integration of virtual network world and the physical world. Resources(资源)(zī yuán), information, physics and people can be closely integrated(整合)(zhěng hé) by CPIS, hence, the Internet of things and related services can be established and the manufacturing plants can be transformed into a smart environment.

2. Cloud computing

In the Internet virtual brain architecture(体系结构)(tǐ xì jié gòu), cloud computing is the Internet virtual brain central nervous system. "Cloud" is actually a metaphor of the network and Internet, it is a pay per use model, this model provides available, convenient, on-demand network access into shared pools of configurable computer system resources and higher-level services that can be rapidly provisioned with minimal management effort, often over the Internet. Google cloud computing has about 1,000,000 servers, Amazon,

IBM, Microsoft and Yahoo companies such as the "cloud" has hundreds of thousands of servers, and cloud computing supports users using a variety of terminal services in any position.

Distributed computing(分布式计算) and virtualization technology(虚拟化技术)are important parts of cloud computing. From the technical point of view, cloud computing provides a new way to manage and use computing resources. For example, under the mode of traditional manufacturing enterprises(企业), to realize the information application, IT hardware and software must be purchased and specialized IT departments established to implement all platforms including IT infrastructure and service system. It not only requires long periods but also needs maintenance personnel to ensure normal operation of the whole system. Cloud computing allows enterprises to get their applications up and running faster, with improved manageability and less maintenance, and that it enables IT teams to more rapidly adjust resources to meet fluctuating and unpredictable demand. Cloud providers typically use a "pay-as-you-go" model, which allows companies to avoid or minimize up-front IT infrastructure(基础设施) costs.

From the user's point, to ensure a safe and reliable application system not concerning with the implementation of the business system, cloud can provide the best solution for its IT infrastructure, development environment and applications, just by acquiring the application system through the client access.

3. Industrial data

Ma Yun has said that today's Alibaba has essentially become a data company, Taobao's purpose is not to sell goods, but access to all retail and manufacturing data. Similarly, Baidu, Tencent and other Internet giants have upgraded their companys' big data strategy. The industrial data in the industrial data collection and data analysis features based on equipment, equipment quality and production efficiency, and optimize the industrial chain management more effective, and for the future manufacturing system to build a free environment. It uses industrial sensors, radio frequency identification (RFID), barcode/two-dimensional codes(二维码), industrial automation control system, enterprise resource planning (ERP), computer-aided technology to expand the amount of industrial data.

The industrial data value is reflected in: firstly, the data can be used to understand and solve the visible problems; secondly, to analyze and predict the invisible problems to understand the root cause and avoid visible problems; thirdly, to explore new knowledge and then redefine(重新定义) the problems. The visible or invisible problems can be avoided in the manufacturing system. The data does not directly create value, the real value for the enterprise is the value and action of the information generated after data analysis, all aspects of data through real-time analysis after the timely decision flow chain(适时决策流程链), or content or basis created for the clients according to the value and service.

4. Industrial robot

The industrial robot is a kind of multi-joint or multi-degree-of-freedom machine used in the field of industry. It can perform the work automatically. The industrial robot is composed of three basic parts: the main body, the drive system and the control system. It can accept human command, or run in accordance(一致) with the prearranged(安排) procedures. It can even be based on the principles of AI technology program guidelines.

From a global perspective, at present Europe and Japan are major suppliers of industrial robots, the four industrial robotic company ABB, KUKA, FANUC and Yaskawa account for about 50% of global market share. One of the best robotic companies in China is Shenyang SIASUN Robot, which is the incubator of the Chinese Academy of Sciences, Shenyang, and has been listed on the Shenzhen gem.

At this stage, industrial robots have been widely used in the fields of automobile, electronic and electrical, metal and machinery(机械), robots instead of manual production is the future development trend of manufacturing industry, which is not only the basis for intelligent manufacturing, but also the future of industrial automation, digital and intelligent guarantee. In McKinsey's report, in 2025 the robot market will create 600 billion to 1.2 trillion RMB of output value. In the process of manufacturing reshuffle, there will be more types of new jobs, and human beings must re-examine their ability to find a new position in the new era.

5. 3D printing

3D printing as the representative of the material manufacturing was proposed in 1984, the prototype in 1986, with 30 years of development, is becoming a promising manufacturing technology. The term is being used to encompass a wider variety of additive manufacturing techniques, in which material is joined or solidified under computer control to create a three-dimensional object. The "bottom-up" method does not need the traditional tools, fixtures and a plurality of processing procedures, the device can quickly produce precision parts of any complex shape, so as to realize the "free parts manufacturing". 3D printing solves many complex structural parts forming, and greatly reduces the machining process, shortens the processing cycle, and the more complex the product structure is, the more obvious the advantage is, 3D printing breaks the traditional mode of industrial production, satisfies the needs of the users by the innovation design. In 2014, the United States Local Motors created the world's first 3D printing car Strati, as shown in Figure 5-3. This car costed ＄3,500, the production cycle: 44 hours. Frame, body, seat, console(中控台), dashboard(仪表盘) and hood were 3D printed, whereas the cable, batteries, tires, rims, suspension, engine and electric windshield were made in traditional way. Compared to the traditional twenty thousand parts of the car, it can be described as very simple, streamline bodywork was made of black plastic, while the outer layer was coated with carbon fiber to increase strength. It was powered by the battery(电池组), the maximum speed reached about 64 km/h.

Figure 5-3 3D printing car（http://incar.tw/tag/Strati）

The establishment of big data for the support of the designer platform, combined with 3D printing personalized customization(定制) advantages, to create distributed manufacturing sites all over the world is a great innovation. On the basis of the distributed manufacturing, it is of no use to address the production unit time consumption, and 10 thousand products in 1 manufacturing and processing plant are basically the same in terms of the production capacity.

6. Knowledge work automation

Knowledge work refers to tasks that require specialized knowledge, complex analysis, careful judgment and creative problem-solving skills. The development of the industrial age in many aspects of human's physical strength has put forward higher requirements, forcing people to rely on industrial automation means to compensate for the lack of physical fitness, in order to implement the operation and maintenance of all kinds of large or sophisticated systems; and likewise, the era of knowledge has also presented higher requirements for the human intelligence, people also need to use the knowledge of industrial automation to compensate for the lack of intelligence, and to better complete uncertainty, diversity(多样性) and complexity of the task.

The automation of knowledge work is the organic integration of intelligence, human and automation. Knowledge industry automation is bound to play a central role in the wisdom of the community, intelligent industry, intelligent manufacturing. At present, the main path and technical knowledge of industrial automation intelligent control, AI, machine learning, man-machine interface, data management and automation based on the wisdom of the physical process from automation to virtual space is the key to cultivate(péi yǎng 培养) and develop the knowledge work automation.

7. Industrial network security

The information network is like a beautiful spider web, which can reach the prey, but cannot stand the wind and rain. Industrial control system is mainly the communication between the machines and equipment(shè bèi 设备). The biggest feature(tè diǎn 特点) is the real-time communication within the control unit. Industrial network security emphasizes availability, integrity and confidentiality. The protection of industrial network security should start from the cultivation of consciousness. It is urgent to popularize the knowledge of information security and change the mode of thinking. Information security is viewed in terms of not only the product, but also the service. A comprehensive system of safety emergency plans should be established to respond to attack, system failure or even collapse. A strong and rapid emergency plan and system recovery scheme, can ensure the minimum operation of the original system.

In the manufacturing industry and the trend of global networking, industrial network security will greatly promote the construction of the manufacturing environment and the continued growth of the global manufacturing industry. This requires the combination of management and products, as well as the combination of enterprises and suppliers, to make the industrial network security system be really in the fight against external malicious attacks.

8. Virtual reality

Virtual reality (VR) technology refers to a computer simulation system, which explores an interactive computer-generated experience taking place within a simulated environment.

VR provides a natural interaction through auditory, visual, feeling and even executed multiple sensory scenes. The immersive(chén jìn shì 沉浸式) environment can be similar to the real world or it can be fantastical, creating an experience that is not possible in ordinary physical reality.

In recent years, VR technology has been applied in the manufacturing, petroleum and medical industries for training and simulation of the production assembly process. For example, in Schneider EYESIM immersive training system, a virtual environment is built with the real data from refinery equipment of petroleum and petrochemical industry(shí huà 石化 chǎn yè 产业) based on traditional training system. The operator can enter the virtual factory with 3D glasses (shown in Figure 5-4), and he can work on the device operation in the main control room and use, while operation process is completely true. The system also allows the operator to deliberately have some irregularities in the performance and experience all the consequences caused by illegal operations(违规操作).

Figure 5-4　VR 3D glasses(https://www.vive.com/cn/pro-eye/)

The immersive VR enables people to visit the different places which are usually hard to access in real world, and can observe and experience the extremely special situations with low cost and less risk.

9. Artificial intelligence (AI)

AI is a branch of computer science, which is considered as the one of the three cutting-edge technologies (genetic engineering, nanoscience, AI) in 21th century. It tries to understand the nature of intelligence and produces a new kind of intelligent machine that can react in a similar way to human intelligence. Research in this area includes robotics, language recognition, image recognition, natural language processing and expert systems. AI is not just about robots, which are actually referred to as containers(容器). The robot sometimes looks like humanoid, sometimes not. But AI is effectively a computer inside the robot. If AI is compared as the brain, robot is the body with a variety of sensors.

Whereas humans somehow are good at a single aspect of AI, which is called weak AI. For example, the AI machine AlphaGo could beat a Go world champion, but it only plays Go. Weak AI is everywhere. AI revolution(革命) is a changing process from weak up to strong, and leaping into super. Nick Bostrom from Oxford, a famous philosopher (哲学家) and AI thinker, defined the super as "in almost all areas, it will be much smarter than the most intelligent human brains, including scientific innovation(科学创新), general and social skills".

It can be expected that AI will become the focus in the next 5 years. The technology of AI will be applied in the manufacturing industry, information industry, software engineering industry, bio-medical technology, retailing(零售), automotive industry and other areas.

5.4 A case study of intelligent manufacturing—Siemens Amberg factory

SIEMENS company believes that there are three major demands in the manufacturing, that is, to improve production efficiency, shorten the time to market and increase manufacturing flexibility. Hence the strategies will be focused on three aspects as follows:

The first is the manufacturing execution system (MES, 制造执行系统) which plays a vital role in the industry. The interface between the automation layer and MES is becoming increasingly important and more seamless, and the flexible production(柔性生产) will be realized across the enterprises. All information should be available through real-time visibility and can be used in the production of network links.

The second is a combination of virtual and reality, that is, the integration of the digital product and the real world. It enables the industry to better approach the challenges of the increasingly higher efficiency, shorter product cycles, more frequent reconfiguration.

The third is the cyber physics system (CPS). CPS can combine the physical plant system and the digital world simulation, and ultimately help to realize easy control and adjustment of production process.

Located in Amberg, the eastern of Germany, SIEMENS factory (EWA) has become a model of the future intelligent factory in 25 years, as shown in Figure 5-5. It is the modern factory based on German industry 4.0, whose degree of automation has reached about 75% and production capacity has increased by 8 times. In EWA, 1,150 employees are mainly engaged in operating computers and monitoring production processes. This is a "simple" factory, which realizes not only the whole digitization of the management, product design, development, production and distribution, but also the real-time data(实时数据) association with American R & D center.

In EWA, the degree of matching between logistics automation and information automation in its production process is significantly higher. The production workshop is in the first floor, while the intelligent logistics distribution system(智能物流配送系统) is at the ground. The logistics distribution system can function on the normal distribution plan and the production line as well. When the logistics are running out, technical personnel scans the logistics number then RFID can automatically transmit the information to the central logistics area which in turn transmits to related logistics distribution accurately. It takes only 15 minutes to complete the whole process with the information system being fully automated(信息系统全自动化).

Figure 5-5　Intelligent plant
(http://www.c-cnc.com/news/newsfile/2016/6/6/1548341.shtml)

In EWA, the main production is programmable logic controller (PLC, named Simatic), as well as other industrial automation products, whose categories reach 1,000 types. Currently, in EWA, each production line has achieved more than 1,000 sites of data collection, which is using SIEMENS's own products. In other words, the production process(生产过程)of the Simatic series is controlled by itself, namely "to produce their own". The production process is completely transparent, 1,150 employees can observe not only the real-time production status information, but also real-time online production status report, in which the real production factory and digital factory are synchronous, and being managed by the unified analysis/management tools. The real parameters and the factory production environment will be reflected by the virtual factory, while the real factory will be controlled through the virtual factory.

Experiment A Two-dimensional motion control platform

1. The objective of the experiment

(1) Understand the composition of the feed mechanism of CNC machine tools.
(2) Study the basic electrical components of CNC machine tools.
(3) Learn the concept of interpolation and linear interpolation programming.
(4) Grasp the composition of CNC system.

2. Experimental equipment

(1) A set of the XY platform equipment;
(2) One GT-400-SV card;
(3) An industrial PC;
(4) Matching pen holder;
(5) Several sheets of drawing paper;
(6) Software development platform。

3. Experiment contents

1) Understanding the experimental device

The composition of the feed mechanism of CNC machine tools is shown in Figure A-1. The mechanical structure contains the rail slider, the ball screw nut, the bearing, the coupling, the machine tool body, as shown in Table A-1. Electrical components include servo motor, drives, motors, limit switches, etc., as shown in Table A-2.

Figure A-1　Hardware components of two-dimensional experimental motion system

Experiment A Two-dimensional motion control platform

This experimental system is an open-architecture CNC system, which consists of IPC and the control card (NC embedded PC type). The display corresponds to the NC interface.

Table A-1 The list of mechanical structure of the components

Icons	Name and type	Usage
	Coupling(SPC1)	Connect two shafts (the master and the slave) to rotate together and transmit torque.
	Linear Guides(RSR9)	Support and guide the moving parts for reciprocating linear motion in a given direction.
	Slider(EGH30CA)	Support the moving parts.
	Single-row angular contact ball bearing housings(BK15)	Support the shaft as a supporting base.
	Double-row angular contact ball bearing housings(BF15)	Support the shaft as a supporting base. It can bear a greater axial force than the single-row one and realize the axial positioning.
	Ball screw and nut pair	Determine the coordinates of the table accurately and transform the rotary motion into linear motion.

Table A-2 Electrical components

Icons	Name and type	Usage
	Servo motor (400W ECMA-C20604RS)	A servo motor is a digitally controlled motor that converts electrical energy into mechanical energy for position control. Servomotor is the control motor with the feedback.
	Server driver(MSDA043A1A)	It is part of the servo system, mainly used in the high-accuracy localization system and effective in controlling the servo motor through the position, speed and torque loop.

Icons	Name and type	Usage
	Photoelectric switch(UCX442)	Control the route of mechanical equipment and limit protection.
	Power(250 W, 24 V, 10 A)	Keep the output voltage constant.

continuous

2) Operational procedure

(1) Check the test platform and turn on the power.

(2) Click on the desktop icon "MotorControlBench. exe" . Open the motion control platform experiment software and select the board type shown in Figure A-2, and then click OK.

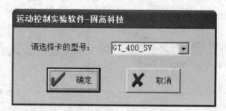

Figure A-2　Board Options Chart

(3) Click the button to enter the Figure A-3 to get two-dimensional interpolation experiment interface.

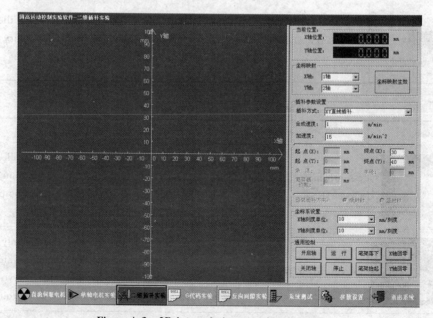

Figure A-3　2D interpolation experiment interface

(4) Enter the speed and the resultant acceleration.

The reference speed is 1 m/min, and the acceleration is 15 m/min^2.

(5) Select XY Plane Linear Interpolation from the drop-down list of "Interpolation" and enter the values for the X end point and the Y end point.

Reference values are shown below:

the end point (X) = 30 mm and the end point (Y) = 40 mm.

(6) Click on [开启轴] to active the servo.

(7) Reset the platform X and Y axes.

The zero-return method is as follows: Click the "X-axis back to zero" button, and the X-axis will return to zero. Until the X-axis returns to the zero, click "Y-axis back to zero".

(8) Fix the test drawing paper, then click the "carriage drop" button to lower the pen holder to draw on the paper.

(9) After confirming that the parameters are set correctly and the axes of the XY platform are reset, click the "Run" button.

(10) Observe the motion of the corresponding click on the XY platform and the motion trajectory of real-time display of the graphic display area in the interface. In the surface of coordinate system settings, the coordinate system scale units for the X and Y axes must be set so that the graphic display is at the appropriate size.

(11) Click the "Carriage Lift" button, lift the pen on the carriage.

(12) Change the motion parameters (synthesis speed, acceleration, end-point positions), repeat steps (2) to (8) and observe the motor motion of the XY table under different motion parameters and the graphics. The experimental data and experimental phenomena should be recorded.

(13) Click on [关闭轴] to turn off the servo motor.

(14) Finish the experiment, shut down.

4. Summary of data analysis and interpretation

According to the experimental results, students are demanded to submit the experimental report and the report should contain the following context:

(1) The interpolation trajectory of the XY platform;

(2) The composition of open CNC system.

Experiment B The measure of positioning accuracy and repeated positioning accuracy to one-dimensional motion mechanism

1. The purpose of the experiment

(1) Understand the composition of the feed mechanical structure of the one-dimensional motion control platform.

(2) Understand the basic electrical components of the one-dimensional motion control platform.

(3) Grasp the concept of positioning accuracy and repeated positioning accuracy, and establish the concept of precision.

(4) Grasp the method to test and calculate the unidirectional accuracy of positioning and repeatability of positioning.

2. Laboratory equipment

(1) One-dimensional platform;

(2) Touch screen;

(3) PLC(S7-200);

(4) Matching steel ruler;

(5) Ultrasonic sensor;

(6) Bluetooth module.

3. Experiment content

1) The experiment devices

The structure of the one-dimensional motion control platform is same to two-dimensional platform, shown in Figure B-1. Inner structure is shown in Figure B-2. The electrical components include 57 Stepper motor, motor driver, limit switch, etc, shown in Table B-1.

Figure B-1　Hardware components of one-dimensional motion platform system

Figure B-2　Inner structure of one-dimensional movement platform

Table B-1　Electrical components

Icon	Name and Model	Effect
	Stepper motor(3A 57BYG)	The motor is an open-loop control motor that converts electric pulse signal into angular or line displacement.
	driver(3ND883)	The stepper motor driver is an actuator that converts electrical pulses into angular displacements. When the stepper driver receives a pulse signal, it drives the stepper motor to rotate a fixed angle (called the "step angle").
	Limit switch(V-156-1C25)	Used to control the travel and limit protection of mechanical equipment.
	Bluetooth module (Risym cc2541)	An integrated Bluetooth PCBA board for short-range wireless communications.

2) Operational procedures

In accordance with GB/T 17421.2—2016 provisions, at least five target positions are selected in the one direction within 1,000 mm. This measuring instrument stroke is 130 mm, so 20 mm is set for each stroke and digital dial indicator is used to measure. Five positions will be set along the axis. The experimental data will be analyzed and compared. Operational procedure of positioning accuracy detection are as follows:

(1) Check the experimental platform, open the electric control panel power switch, power the system.

(2) Click the "back to zero" button on the touch screen or on the mobile phone to let the platform back to zero. The touch screen interface is shown in Figure B-3 and the mobile phone control interface is shown in Figure B-4.

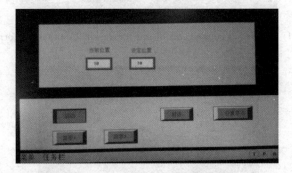

Figure B-3　Touch screen interface

Figure B-4　Mobile phone software interface

(3) Click on the set position, enter "20", the platform forward movement about 20 mm. At this position, the digital dial indicator will show the real distance of the platform movement, such as 20.003. Note that $X_{ij} \uparrow = P_{ij} - P_i = 0.003$ mm.

(4) Repeat the steps (2) and (3) for the same "20" mm travels by the unidirectional.

(5) Repeat the steps (2) and (3) for another four "20" mm travels by the unidirectional.
(6) The accuracy of the platform is calculated by the data obtained.

4. Analysis of positioning accuracy detection experiment data

The symbols " ↑ " and " ↓ " indicate the forward direction or the negative direction.
(1) Calculate the mean unidirectional positional deviation at a position, $\overline{X}_i \uparrow$:

$$\overline{X}_i \uparrow = \frac{1}{n} \sum_{j=1}^{n} X_{ij}$$

(2) Estimate the unidirectional standard uncertainty of positioning at a position, $S_i \uparrow$:

$$S_i \uparrow = \sqrt{\frac{1}{n-1} \sum_{j=1}^{n} (X_{ij} \uparrow - \overline{X}_i \uparrow)^2}$$

(3) Calculate the unidirectional repeatability of positioning at a position, $R_i \uparrow$:

$$R_i \uparrow = 4 S_i \uparrow$$

(4) Get the unidirectional repeatability of positioning at all positions, $R \uparrow$:

$$R \uparrow = \max[R_i \uparrow]$$

(5) Get the unidirectional accuracy of positioning at all positions, $A \uparrow$:

$$A \uparrow = \max[\overline{X}_i \uparrow + 2 S_i \uparrow] - \min[\overline{X}_i \uparrow - 2 S_i \uparrow]$$

5. Summary and writing the report

Recommended book

[1] LI J, NI J, WANG A Z. From Big Data to IM [M]. Shanghai: Shanghai Jiaotong University Press, 2016.
[2] CHEN M, LIANG N M. Numerical factory of IM road [M]. Beijing: Mechanical Industry Press, 2017.
[3] Made in China 2025 Plan Unveiled[J]. Making paper information, 2016(8): 93.
[4] XIA Y N, ZHAO S. China Manufacturing [M]. Beijing: Mechanical Industry Press, 2016.
[5] Olexa R. The Father of the Second Industrial Revolution [J]. Manufacturing Engineering, 2001, (2): 127.
[6] SINUMERIK 840Dsl[EB/OL]. (2019). http://www.industry.siemens.com.cn/automation/cn/zh/automation-systems/cnc-systems/Pages/Default.aspx.
[7] Fanuc i series CNC[EB/OL]. (2018-01). http://bj-fanuc.com.cn/HomePage/HomePage.
[8] PC104[EB/OL]. (2014-04-17). http://www.deltatau-china.com/.
[9] Vertical machine center [EB/OL]. (2018-01). https://www.haascnc.com/index.html.
[10] LI E L. Interpolation theory of CNC system[M]. Beijing: National Defense Industry Press, 2008.
[11] Iron Mu Zhou Gantry Machine Center, Changsha, Hunan [EB/OL]. (2017-12-19). http://b2b.liebiao.com/jixiejihangyeshebei/438315111.html.
[12] RAO Z G and WANG Y W. Ball Screw Pair Transmission and Self-locking Device [M]. Beijing: National Defense Industry Press, 1990.
[13] HU Q B, HU H B and LV Z Y. Fully Digital Servo Control System Based on AC Permanent Magnet Synchronous Motor[J]. Power Technology Application, 2004,(2): 69-72.
[14] WANG Y H. Quasistop Control and Application of CNC Machine Tool Spindle[J]. Machine Tool Electrical Apparatus,2009: 15-16.
[15] Ribs[EB/OL].(2018-05). https://en.wikipedia.org/wiki/Longeron.
[16] Gantry machine center[EB/OL].(2015-08).http://www.haitecnc.com/case/longmen/6.html.
[17] Thermal deformation[EB/OL]. (2019). https://www.heidenhain.com/en_US/products/linear-encoders/.